Genealogies *of* Environmentalism

Genealogies *of* Environmentalism

The Lost Works *of* Clarence Glacken

EDITED BY S. RAVI RAJAN
WITH ADAM ROMERO AND MICHAEL WATTS

University of Virginia Press
Charlottesville and London

University of Virginia Press
© 2017 by the Estate of Clarence J. Glacken
All rights reserved
Printed in the United States of America on acid-free paper

First published 2017

1 3 5 7 9 8 6 4 2

LIBRARY OF CONGRESS CATALOGING-IN-PUBLICATION DATA

Names: Glacken, Clarence J., author. | Rajan, S. Ravi, editor. | Romero, Adam, editor. | Watts, Michael, 1951– editor.
Title: Genealogies of environmentalism : the lost works of Clarence Glacken / edited by S. Ravi Rajan with Adam Romero and Michael Watts.
Description: Charlottesville : University of Virginia Press, [2016] | Includes bibliographical references and index.
Identifiers: LCCN 2016002831| ISBN 9780813939070 (cloth : alk. paper) | ISBN 9780813939087 (pbk. : alk. paper) | ISBN 9780813939094 (e-book)
Subjects: LCSH: Natural history—History. | Environmentalism—History. | Naturalists—Biography. | Environmentalists—Biography.
Classification: LCC QH15 .G53 2016 | DDC 508—dc23
LC record available at http://lccn.loc.gov/2016002831

Contents

Foreword   vii
    Michael Watts

Preface   xv

Introduction   xix

Progress, Population, and Nature from the Age of Reason to the Mid-Nineteenth Century   1

The Aesthetic and Subjective Appreciation of Nature: Alexander von Humboldt   21

Darwin and His Contemporaries   31

In the Wake of Marsh   95

Key Trends in Nineteenth-Century Environmentalism   133

The Malthusian Shadow over the Twentieth Century   166

A Selected Bibliography of Clarence Glacken's Works   197

Index   199

Foreword

> Large related bodies of thought thus appear, at first like distant riders stirring up modest dust clouds, who, when they arrive, reproach one for his slowness in recognizing their numbers, strength and vitality.
> —Clarence Glacken, *Traces on the Rhodian Shore*

By some strange twist of fate, I occupy Clarence Glacken's old office in the Department of Geography on the fifth floor of the McCone Building on the Berkeley campus. Glacken's old office faces east, looking up into the Berkeley hills and beyond, to the eucalyptus near the Lawrence Hall of Science. I came to Berkeley in the fall of 1979. Glacken was then retired—he had stepped down from full-time teaching following a heart attack in 1974—but was still active, working assiduously every day on the sequel to *Traces on the Rhodian Shore*. Clarence was a striking figure: tall and ramrod straight, with, at that point in his life, long gray hair, often pulled back into a ponytail. He was neither physically nor intellectually imposing: quite the contrary, he was quiet and sensitive. Modesty seemed to me to be his distinguishing trait. Clarence always seemed to me a model of Old World gentility; he was gentle to a fault.

Clarence James Glacken (1909–89) was a third-generation Sacramentan. His paternal grandmother had moved to the city in a covered wagon as an infant in 1854, and both grandfathers settled in the city during the 1870s. His parents were Sacramento natives, born and raised downtown. Growing up at 1830 T Street with his younger brother, Glacken counted among his immediate family his mother (his parents had divorced during World War I), an uncle, and maternal grandparents.[1]

---

1. An excellent short biography was published by Corinna Fish, a former Berkeley geographer, in the *Sacramento Bee*, January 30, 2012 (http://sacramentopress.com/2012/01/30/traces-of-a-native-son-searching-for-clarence-glacken/).

In his own account, Glacken's historical and geographic sensibility was forged in northern California. It was the gold rush, the Central Pacific Railroad, and the confluence of the Sacramento and American rivers—and his stamp collection!—that sparked his passion for landscape and their deep histories. Looking back in later life on the earliest years, he said, "One must be careful to avoid a teleological view... there were, however, several interests in early life which I later perceived to be geographical and historical. Sacramento... is an historic city, perhaps not by Old World standards, but certainly by American ones."[2]

Glacken attended Sacramento Junior College for two years, where he was encouraged by his English instructor to study with the great historian Frederick Teggart at the University of California, Berkeley. It was Teggart who became—and in many senses remained—the towering intellectual figure in Glacken's life. In his first year, in addition to a course on the politics and history of the Far East, Glacken enrolled in Teggart's course, "The Idea of Progress." The undergraduate class dealt with cyclical theories and myths of the golden age in the ancient world, with providential interpretations of history in the Middle Ages, and with the idea of progress in the modern era. Teggart (1870–1946) had begun his career as a librarian and archivist, examining the problems of cataloguing ancient libraries and working on specialized bibliographies on railways, oil, forests, or Victorian poetry. In 1916 he published a book on the relationship of history to other branches of knowledge (*Prolegomena to History: The Relation of History to Literature, Philosophy and Science*, University of California Press, 1916), and two years later he published *The Processes of History* (Yale University Press, 1918). By the time Glacken appeared in his class it was the idea of progress that compelled him. During his years at Berkeley, Glacken "had no time for anyone but for Teggart," who was "a revelation." Alfred L. Kroeber and Robert H. Lowie—two other towering figures on campus—and Carl O. Sauer were not part of his intellectual formation. Glacken was aware of Sauer, but his sole interest lay in Teggart. He devoted himself almost exclusively to the study of the history of ideas: certainly to the ideas of Teggart himself, but also to those of the likes of Arthur Burtt and his book, *The Metaphysical Foundations of Modern Physical Science* (1925), and the scholarship of John Livingston Jones on the English poet Coleridge. Glacken received his undergraduate degree with highest hon-

---

2. All quotations from Glacken are taken from his "A Late Arrival in Academia," in *The Practice of Geography*, ed. Anne Buttimer, 20–34 (London: Longmans, 1983). Other useful references are Clarence Glacken, *Traces on the Rhodian Shore* (Berkeley: University of California Press, 1967); idem, "Man against Nature: An Outmoded Concept," in *The Environmental Crisis*, ed. H. W. Helrich, Jr., 127–42. (New Haven: Yale University Press, 1970; repr., Warner Modula Publications, 1972); and idem, "Man and Nature in Recent Western Thought," in *This Little Planet*, ed. Michael Hamilton, 163–201 (New York: Scribner's, 1970).

ors in 1930 and his master's degree in 1931, both from the Teggart-chaired Department of Social Institutions, an innovative and unusual program that long preceded other much-heralded multidisciplinary initiatives, such as Harvard's Department of Social Relations, which were to appear after World War II.

Under Teggart's tutelage, Glacken did not, of course, escape geographic ideas. Among the teachers in the Department of Social Institutions was a young assistant, Margaret Hodgen (1990–77), who at the time was working on the history of anthropology and the dynamics of social change. Through her class, Glacken deepened his understanding of not just the anthropology classics but also the history of environmental ideas, including works by French geographers and the American geographer Ellsworth Huntington. Hodgen was part of an anthropological community in which the relations between the discipline, space, and environment—one thinks of culture areas and the early iterations of what was to become cultural ecology—were developing apace in and around Kroeber and Sauer. Yet none of this seemed to push Glacken toward geography.

Glacken's route back to Berkeley as a faculty member traced a circuitous, two-decade-long circumambulation of the globe. Glacken's graduation from Berkeley launched him into what he called the "bleak years" of the Great Depression. In the five years following his departure from Berkeley, Glacken worked for the newly established Farm Security Administration (FSA), reporting on migrant labor camp conditions throughout the state of California, from Redding to Bakersfield. With the savings he could muster from his FSA position, in 1937 Glacken traveled from San Francisco to Japan, then to China, Indochina, the Middle East, the Mediterranean, and Europe.[3] Without this "species of fieldwork," Glacken once claimed, he would neither have come to grasp the full meanings of cultural difference nor have been able to move beyond "a world of abstractions."

In 1941, three months after his first wife died, Glacken was drafted into the army and served for six years. During his service he married Mildred Mosher, who was an assistant to the influential Malthusian ecologist William Vogt, author of the bestseller *Road to Survival* (1948) and associate director of the Division of Science and Education of the Office of the Coordinator of Inter-American Affairs. Following Japan's surrender and the emergence of a "northeast arc" strategy, the U.S. government quickly recruited and trained specialists in Japan and Korea to staff management positions in those territories, and Glacken was selected as an officer-student for training in Japanese affairs at the Civil Affairs Training School, which had been formed at the University of Chicago and was directed by the anthropologist Fred

---

3. David Hooson, "In Memoriam: Clarence Glacken 1909–1989," *Annals of the Association of American Geographers* 81, no. 1 (1991): 152–58.

Eggan. He studied Japanese, and his teachers included prominent orientalists, among them John Embree. Subsequently Glacken was sent by the army to Korea, where he acted as director of the Office of Health and Welfare in the military government and subsequently with the Veterans Administration.[4]

At the age of forty, Glacken entered the Isaiah Bowman School of Geography at Johns Hopkins University. His dissertation, "The Idea of the Habitable World," examined how the history of ideas could be drawn on to reframe contemporary debates about population growth and natural resources. As soon as he had completed his PhD, in 1951 (in record time), he was appointed to a short-term position to conduct an ethnographic study of Okinawa for the Pacific Science Board of the National Research Council. He studied three villages in Okinawa, focusing on the family system, land tenure, and the impact of land tenure on the organization of space. This work resulted in a book, *The Great Loochoo* (1956). After returning to his family in California, he visited Carl Sauer (how and why this transpired is unclear), who promptly offered him a position at Berkeley in the fall of 1952.

In some respects, much of what passed as the Berkeley school in the 1950s was clearly orthogonal to the sorts of questions in which Glacken was interested. Glacken respected Sauer (and vice versa), but it is not clear that they were close, and they certainly had very different personalities.[5] Yet Glacken arrived at a propitious moment in Sauer's own intellectual evolution. He had become despondent about much of what passed as American geography and about the emerging spatial revolution. Sauer took pride on being on the West Coast, remote from the midwestern and northeastern centers of academic geography, which he found repellent and considered second rate. Not least, his deep skepticism about much of what passed for development converged with a growing and wide-ranging set of concerns that he had come to appreciate (shaped by the likes of Lewis Mumford, Kenneth Boulding, Edgar Anderson, and others) on the state of the Earth. Here Sauer drew Glacken into the famous 1955 international symposium at Princeton University, "Man's Role in Changing the Face of the Earth" (organized by Sauer, Marston Bates, and Lewis Mumford), the first academic conference on anthropogenic environmental change, seven years prior to the publication of Rachel Carson's *Silent Spring* and fifteen years before the first Earth Day. At that conference, Glacken presented a paper on the changing ideas about the living world, summarizing some themes of his doctoral thesis even as he clarified his own trio of ideas, which were to become the backbone of *Traces:* the idea of a divinely designed Earth (both eco-

---

4. Anne Macpherson, "Clarence James Glacken 1909–1989," in *Geographers: Biobibliographical Studies,* ed. Geoffrey J. Martin, 27–42 (London: Mansell, 1992).

5. M. Williams, D. Lowenthal and W. Denevan, *To Pass On a Good Earth* (Charlottesville: University of Virginia Press, 2014).

logical theory and the intelligent design argument are direct descendants), the idea of environmental influence on people, and the idea of human influence on the environment. This exposure and the clear political implications of examining the dramatic human impact on the Earth—the forerunner, in short, of what we now call the Anthropocene—drew Glacken further into the emerging environmentalism of the late 1950s and 1960s. Other invitations followed: to a 1961 conference at the Pacific Science Congress organized by Ray Forsberg and titled "The Role of Men in Island Ecosystems," and to another, in April 1965, organized by F. Fraser Darling and John Milton and titled "The Future Environment of North America."

From the time of his arrival at Berkeley, Glacken was inevitably caught up in and reluctantly drawn into this nascent environmentalism. It had the effect of promoting and endorsing his sophisticated historical approach to human transformations of the Earth, while all the time expanding the scope of his project. Glacken's colleague, David Hooson, wrote that *Traces* "was originally planned as an introductory chapter to a major work on these themes in 19th and 20th century thought. . . . [*Traces*], however, stands on its own as one of the most scholarly books written by a geographer, or by a historian of ideas, in this century."[6] That introductory chapter finally checked in at more than 700 pages. At the same time that he was writing *Traces,* Glacken was simultaneously researching environmental thought of the nineteenth and twentieth centuries, and actually began writing the sequel as soon as *Traces* went to press. Paradoxically, Glacken himself was not a political creature. He shied away from the political implications of his scholarship. Yet within a decade of his return to Berkeley he was projected into politics of a most fearsome sort in the 1960s, and in many respects it destroyed him.

In keeping with the pace and rhythm of Glacken's entire life, *Traces* had a slow gestation, appearing fifteen years after he took up his Berkeley appointment (and three years after he was awarded tenure). It was, of course, widely acclaimed, but did not sell terribly well. Its intellectual capaciousness and erudition aside, *Traces* was a tough read for the uninitiated. The book clearly bears the imprint of an earlier intellectual generation, of Arthur O. Lovejoy, Louis B. Wright, J. Shapiro Salwyn, and Glacken's own teacher, Frederick J. Teggart. In this sense the book was somewhat unfashionable. Its theoretical referents are quite out of keeping with the sorts of approaches to the sociology of knowledge production and conceptual history that were to come later.[7]

The 1960s turned out to be a bittersweet period in Glacken's life. He was now tenured (at the ripe old age of fifty-five), had become chair of the department,

---

6. Hooson, "In Memoriam: Clarence Glacken," 156.
7. Raymond Williams, *Keywords* (London: Oxford University Press, 1976).

had been awarded a Guggenheim Fellowship, and had gradually built up a raft of courses (on ideas about nature, and on cultural landscapes) capable of the same sort of appeal as the courses taught by his old teacher, Teggart. Perhaps most important for Glacken's own scholarship, he now had as a colleague Paul Wheatley, newly arrived from Cambridge University, a person who could match him step for step as a multilingual and cross-cultural intellect of the highest rank. Wheatley was an extraordinary figure[8]—a student of H. C. Darby at Cambridge, he left Berkeley in 1969 to return to University College London, then promptly moved to the University of Chicago, where he chaired the famed Committee on Social Thought—and he and Glacken jointly taught much-lauded courses on Asia. Both Glacken and Wheatley were exceptionally gifted teachers, if very different in style and temperament, and their courses became legendary. The future looked rosy.

But it was not to be. Glacken's professional success coincided with a period of intense political turmoil, both within the field of geography (the spatial and quantitative revolution was a major assault on everything the Berkeley school stood for) and on the Berkeley campus. As the United States descended into the nightmare of Vietnam, the campus exploded in acrimony and anger over American empire, and the academy became politicized (and contentious) in ways that made the 1964 free speech movement seem calm by comparison. Protests, occupations, police violence, and hostility expressed toward the campus by Ronald Reagan as he assumed the governorship of the state made the university close to ungovernable. It fell to Glacken, whose personality was ill suited for coping with, much less managing, the battles raging within his department and on campus, to be departmental chair during this period. Two years of struggling with escalating faculty rivalry and student protests took their toll. Glacken had a nervous breakdown in the spring of 1970 and a physical breakdown the following fall. Following a six-month leave of absence, Glacken resumed his duties and showed signs of slow recuperation over the next few years, but the path to normalcy was tortuous.

When I arrived in 1979, Clarence was in fine fettle. I vividly recall that during one of our regular Sunday discussions he asked me what I was writing about. I gave a terse, deliberately vague response. The triviality of whatever it was I was working on seemed too painful to rehearse. "Why don't you give me something to read?," he said. "We can talk about it next week." Reluctantly, I handed over some bromide, hoping against odds that Clarence would forget about it. But sure enough, the following week he came by *my* office—barely larger than a broom closet, though with

---

8. Paul Wheatley. *Pivot of the Four Quarters* (Edinburgh: University of Edinburgh Press, 1972); idem, *The Places Where Men Pray Together: Cities in Islamic Lands, Seventh through the Tenth Centuries* (Chicago: University of Chicago Press, 2001).

a magnificent view of the Bay—to talk. In his characteristically generous way, he sang the praises of my work—the style, the argument, the sensitivity to history were first class—all delivered with Clarence's customary gentleness. Then there was silence. A long silence. But, he said finally, we *have* known this for about a thousand years.

In 1982, as far as it is possible to reconstruct, Clarence delivered the sequel to *Traces*, some two thousand pages long, to the University of California Press. From here on the story is confused and incomplete. The manuscript appears to have landed on a junior editor's desk, and by whatever strange sequence of events—it is not even clear that the manuscript was sent out for review, nor, apparently, was a Xerox copy made by the press—the manuscript was returned. There is no record of what exactly transpired and what message was delivered to Clarence. Was it simply a question of length and of cutting it down? Was the book simply out of fashion—and outside the prevailing "market model" of short, commercially oriented books? What we do know is that it was personally devastating for Clarence. Once more Glacken descended into a period of great darkness. Only later did it come to light that he had apparently destroyed all remaining copies of the manuscript.

Glacken's own psychological and emotional deterioration in the wake of the manuscript's completion is deeply saddening. Mildred, his second wife, had died of a stroke in 1980. Her passing represented a terrible loss, for obvious reasons, and the removal of a key anchor in his life. By the mid-1980s he was clearly unwell and had come to doubt the value of his entire corpus.

Almost a half century after its publication, *Traces on the Rhodian Shore* remains a towering achievement, and Glacken's scholarship seems more relevant than ever if we are to place the Anthropocene on a larger landscape of the history of human ideas and practice. I can do no better than conclude with a quotation from Corinna Fish's hommage to Glacken, published, appropriately, in the *Sacramento Bee:* "The ideological and intellectual parameters of contemporary debates, from anthropogenic climate change to fracking, can be traced directly to the three ideas Glacken recognized and analyzed like no other scholar before or since."

Michael J. Watts

Preface

Clarence Glacken's *Traces on the Rhodian Shore* is a classic in the history of geographic and environmental ideas. First published in 1967, it remains in print today, almost fifty years later. Glacken wrote a number of other papers, but for some reason, about which there is more speculation than hard fact, he never published the sequel to *Traces,* although many independent accounts confirm that such a sequel once existed in its entirety.

About two decades ago, while working in the archives at the Bancroft Library at the University of California, Berkeley, I stumbled upon a neatly handwritten manuscript that, barring footnotes, appeared quite complete. Further research indicated that this was not Glacken's much vaunted sequel but a totally different book manuscript, one he had proposed to publishers in the aftermath of the publication of *Traces.* Subsequently some of Glacken's colleagues, notably Professors David Hooson and David Stoddart, shared printed draft papers that Glacken had given them. In the following years I conversed with a number of his former students and colleagues in a bid to learn more about Glacken and his works since the publication of *Traces.* During this period I was extremely fortunate to enjoy the support of Professor Michael Watts, who also provided many valuable leads. Together, we decided to find a way to publish what remained of Glacken's lost works.

Finding a publisher proved difficult, given the tight budgetary times and the consequent reluctance on the part of publishers to take on edited works. I therefore decided to approach this project differently and produce the book as a coherent whole, rather than a disparate collection of essays. Here I have made a conscious editorial choice. As an organizing principle, I decided to follow a historical chronology, interspersing essays from both sources, the printed papers that Stoddart and Hooson gave me with those from the handwritten manuscript. In essence, all essays, representing key thinkers and their writings, are arranged by the dates that the protagonists lived and worked. The result is this book, which carries forward the key themes from *Traces* into the eighteenth, nineteenth, and twentieth centuries.

Even while I was shopping the book around to potential publishers, Adam Romero, then a brilliant graduate student in the Department of Geography at UC Berkeley, had independently developed an interest in Glacken. We decided to collaborate, and in the ensuing months, Adam painstakingly typed out the entire handwritten manuscript, carefully checking his typescript against the original source. But for his herculean effort, this book would not have been possible. Last year the project finally acquired momentum. The University of Virginia Press decided to acquire it, and has been an enthusiastic backer.

Editing the manuscript for publication, however, required considerable effort. Because both sources, the handwritten manuscript and the typescript, were drafts, many passages lacked coherence. There were no footnotes in the handwritten script, and in the draft papers he shared with Hooson and Stoddart, the notes were often incomplete and in many instances inaccurate. Many historical figures were mentioned in the text, but without any indication of who they were. There were also a number of key concepts that were not defined or explained. None of this was Glacken's fault, for these were but first drafts. In approaching the editorial task, I adopted three central principles: fidelity to the original texts, readability and coherence, and adequate footnoting to assist the reader. Accordingly, I have removed several passages that appeared not to make sense. I also cut or paraphrased many large quotations. Glacken, in these first drafts, had many such quotations, some several pages long, and he very likely intended to edit them himself at a later stage. In editing the lengthy quotations my goal was to be least disruptive, to retain the parts that added value to his argument while removing those that were repetitive. I also tried my best to verify Glacken's sources. Here I have only partially succeeded. There remain some quotations, especially in languages I do not read, that I could not find an exact page reference to, although in most instances I have managed to locate Glacken's source. Next, for every name that Glacken mentions, I have provided a brief biographical statement to explain who the person was. Likewise, for every keyword or concept I have ensured there is at least a brief definition. Last but by no means least, since Glacken did not provide introductions to the draft essays, I wrote an editor's introduction for each essay, summarizing the key themes. There is also an introduction to the volume at large.

One important caveat is in order here. This book contains only Glacken's unpublished work. There are other works, published as journal articles or book chapters, that could have rounded out the content even more and made the chronology more complete. There was therefore a definite case for including this material. However, there was also a trade-off. Given the word limit agreed to with the publisher and the already generous length of the entire selection, I had to choose either to augment the text by adding footnotes to substantiate the text in the manner described earlier or leave it as it was and include more content. In the end, I decided to adopt the former strategy, in a bid to make this book more useful and leave a more lasting legacy. However, for interested readers, a bibliography of Glacken's publications is provided at the end of the book.

This project would not have been possible without the assistance of several people. Without doubt, my two collaborators have been critical to the project. Adam Romero was painstaking in transcribing the handwritten manuscript and subsequently reading through and editing the draft, and Michael Watts, an immense source of strength and support. This book would also never have seen the light of day but for David Hooson. It was his kindness, generosity, and good humor, as well as his countless stories about Glacken, that convinced me in the first instance that this project needed to be undertaken. I am also immensely thankful to David Stoddart for preserving the papers that Glacken gave him and allowing us to use them here. Thanks are owed to many other of Glacken's colleagues and students. Among them are Carolyn Cartier, who shared Clarence Glacken's course readers, and James Parsons, Alan Pred, Richard Walker, Victor Savage, Diana Davis, Kären Wigen, Martin Lewis, James Sideway, and Terry Burke. I am extremely grateful to the staff and faculty of the Department of Geography at UC Berkeley and to the librarians and archivists of the Bancroft Library. Last but not least, the world owes a huge debt of gratitude to Mary Elizabeth Braun of Oregon State University Press, a geographer herself, for recommending publication to her colleagues at the University of Virginia Press; to two amazingly insightful editors there, Boyd Zenner and Ellen Satrom; to the thorough and often insightful copy editor, Marjorie Pannell; and to the three referees the press sent the manuscript to for review. That said, any mistakes here are entirely mine.

S. Ravi Rajan

Introduction

Long before environmental history existed as a scholarly field of inquiry, a geographer at Berkeley wrote a book that would come to be one of its foundational texts. In 1967, Clarence Glacken published *Traces on the Rhodian Shore,* a big history of ideas about the environment from antiquity to the eighteenth century.[1] The breadth of scholarship and the scope of the book were breathtaking, its erudition and command of the history of ideas without peer.

The book initially received mixed reviews. For the most part, the detractors were disciplinary practitioners who questioned either a detail or the method from the standpoint of their fields. On the other hand, many scholars praised *Traces* for providing a complex intellectual history that demanded multidisciplinary and multilingual expertise. By the time the paperback edition arrived, any debate about its place in the pantheon of great books in environmental history had been put to rest. A reviewer in the journal *Professional Geographer,* for example, called it "one of the best and most important books published by a geographer in the English-speaking world in the last hundred years," and one for the *American Anthropologist* wrote that "a book such as this rarely appears anymore."[2] Today, *Traces* is essential reading for anyone interested in environmental ideas and their history. As J. R. McNeil, past president of the American Society for Environmental History, put it, "For the Western intellectual tradition up to the eighteenth century, the most comprehensive and insightful text remains Clarence Glacken's *Traces on the Rhodian Shore.*"[3]

For the benefit of readers who might not have encountered Clarence Glacken before, it might be useful to sketch a brief biography. Glacken was a third-

---

1. Clarence J. Glacken, *Traces on the Rhodian Shore: Nature and Culture in Western Thought from Ancient Times to the End of the Eighteenth Century* (Berkeley: University of California Press, 1967).
2. For a detailed summary of the various reviews, see S. Ravi Rajan, "Clarence Glacken: Pioneer Environmental Historian," *Environment and History,* in press.
3. J. R. McNeil, "The Nature of Environmental History," *History and Theory* 42 (2003): 5–43.

generation Sacramentan, born in the city in 1909. He was the son of two California families. His paternal grandmother migrated to California in a covered wagon as an infant in 1854; his maternal grandmother was born in Sheldon (near Elk Grove) in 1865.[4] Both grandfathers had settled in the state by the 1870s, and both his parents were born and raised in downtown Sacramento. As a child, he was enthralled, as many children are, by philately and the mystique of stamps from far-off places. In a 1983 autobiographical essay, Glacken characterized his California upbringing as central to the interests he developed in *Traces:* "Looking back in later life on the earliest years, one must be careful to avoid a teleological view.... There were, however, several interests in early life which I later perceived to be geographical and historical. Sacramento ... is an historic city, perhaps not by Old World standards, but certainly by American ones."[5] In this account it was the gold rush, streetcars, the Central Pacific Railroad, the State Capitol, and the confluence of the Sacramento and American Rivers that triggered his passion in the intersection of geography and history. By high school, Glacken had developed a strong interest in geography, and this interest was harnessed by a young English teacher named John Harold Swan, who had recently graduated from Berkeley.

Glacken attended Sacramento Junior College, now Sacramento City College, for two years, before transferring to UC Berkeley. On Swann's recommendation, he decided to study in a department named Social Institutions, which had been founded by the historian Frederick Teggart. The latter was an autodidactic Irish scholar who wrote *Rome and China,* one of the first books in the field that is known today as world history.[6] Glacken was greatly influenced by a yearlong lecture course taught by Teggart. It was an encyclopedic course, covering human history, demography, psychology, historical geography, and the methods of the social sciences. (Another formative figure for early Glacken was Margaret Hodson, who lectured on the history of social thought and social theory and introduced him to the possibilist tradition in geographic thought.) For Glacken, the Teggart course was eye-opening: "Teggart's course, The Idea of Progress ... was a revelation, because I then realized the importance of the history of ideas.... I have often been asked whether in these undergraduate years I took any geography courses. I did not. I had heard of Carl Sauer [Glacken's future boss], but I had no time for anybody but Teggart."[7]

Glacken received his bachelor's degree with highest honors in 1930 and his mas-

---

4. Corinna Fish, "Traces of a Native Son," *Sacramento Press,* January 30, 2012.
5. Clarence J. Glacken, "A Late Arrival in Academia," in *The Practice of Geography,* ed. Anne Buttimer, 20–34 (London: Longman's, 1983).
6. Frederick J. Teggart, *Rome and China: Study of Correlations in Historical Events* (Berkeley: University of California Press, 1939).
7. Glacken, "A Late Arrival in Academia."

ter's degree in 1931, both from the Department of Social Institutions. By this time he was fluent in German, French, and Spanish, and well versed in the classical languages as well. Over the course of his career he learned Norwegian, Danish, Swedish, and Japanese.

The years after he graduated did not afford many opportunities for a young graduate from a fledgling program called Social Institutions. The Great Depression obviously limited his options. By his own account, the early 1930s were "bleak years for young people."[8] Given the absence of scholarships, Glacken joined the Farm Security Administration, set up during the first Rooseveltian government. After years on the road conducting surveys in refugee camps from Redding to Bakersfield, Glacken resigned in 1937 and embarked on a world tour. As he put it, "In retrospect, I look upon my travels as a species of field work.... I do not think I would have ever developed my intense enthusiasm for the history of ideas without it. It would have been a world of abstractions."[9] Following his global tour, he returned to the Farm Security Administration for an additional four years. In the introduction to *Traces*, he wrote, "The early stimulus to study these ideas came also from personal experience.... As I worked with resident and transient families on relief, with migratory farm workers who had come from the Dust Bowl, I became aware... of the interrelationships existing between the Depression, soil erosion, and the vast migration to California."[10] He was drafted into the army in 1941, after the death of his first wife, and served in the Far East. After the war he was posted to Japan and Korea. He was trained by the army as a specialist in Japanese language and culture. Upon discharge, he accepted a job in Washington with the Veterans Administration. During this period of service he married again and had two children.

It became clear to Glacken that his passions lay in the history and geography of ideas, and so, in 1949, at the age of forty, he entered graduate school at Johns Hopkins University. He wrote a dissertation on the relation of population growth to resources, the influence of the geographic environment on human culture, the contrasting idea of man as a geographic or a geological agent, soil concepts, and living nature as a series of complex interrelationships. In each instance he traced Western attitudes from the mid-eighteenth century to the mid-twentieth.[11]

On completing his thesis at Johns Hopkins, Glacken joined the Geography Department at UC Berkeley after a short interview with the then chair, Carl Sauer. He tried publishing his thesis but was rebuffed by publishers on the grounds that

8. Ibid.
9. Ibid.
10. Glacken, *Traces on the Rhodian Shore*.
11. Clarence Glacken, "The Idea of the Habitable World" (PhD diss., Johns Hopkins University, 1951).

any book on the environment that had been successful was in the nature of a crusade, and that a historical approach to the subject would not attract many readers.¹² He then turned to publishing other things, starting with a book drawing on his work and subsequent fieldwork in the Ryukyu Islands and on Okinawa.¹³ He also began teaching a course titled "Relations between Nature and Culture." The course, which anchored much of Glacken's future research interests, had three main themes: (1) environmental determinism, (2) environmental change induced through human agency, and (3) the teleological idea of nature, and the concept of a designed Earth. Glacken also participated, along with Lewis Mumford, Marston Bates, and Carl Sauer, in the landmark 1955 Princeton international symposium, "Man's Role in Changing the Face of the Earth." His contribution, titled "Changing Ideas of the Habitable World," was subsequently published in a famous volume bearing the same title as the conference. He then wrote a paper titled "This Growing Second World within the World of Nature," which he presented at the Pacific Science Congress in Honolulu in 1961, and a paper on the eighteenth-century French naturalist, Count Buffon, among other topics.[14]

Then began a long journey, lasting several years and requiring endless correspondence with many publishers, that finally resulted in the publication of *Traces*.[15] The book had three main themes, echoing those of his course. They were:

a) The idea of a designed Earth, an example of the doctrine of final causes as applied to the natural processes on Earth, which, Glacken argued, underlay the emergence of natural history and ecological theory and its interpretation of the nature of earthly environments as wholes and as manifestations of order;

b) The idea of environmental influence, which, building on the ancient contrast between *physis* and *nomos*, that is, between nature and law or custom, led first to climate being held responsible for the inebriety or sobriety of entire peoples in the post-Reformation period and then to the idea of the environment as a whole imposing limitations on all life, as adumbrated in the work of such thinkers as Montesquieu, Wallace, Hume, and Malthus; and

c) The idea of man as a modifier of nature, according to which man is seen as

---

12. Rajan, "Clarence Glacken."

13. C. Glacken, *The Great Loochoo: A Study of Okinawan Village Life* (Berkeley: University of California Press, 1955).

14. For complete citations, please refer to the bibliography at the end of this book.

15. For an account of this fraught correspondence, see Rajan, "Clarence Glacken."

finishing nature. Glacken argued that whereas in the work of people like Ray and Buffon, this role for man was seen as positive and optimistic, by the mid-nineteenth century, especially in the work of Marsh, it was seen as negative and pessimistic.

For a time after the publication of *Traces,* Glacken pursued a range of intellectually connected projects that spoke to the themes he had addressed in the book. He applied for a Fulbright Fellowship in 1956 to work in Norway, expecting to prepare a case study of that country as an illustration of "the relationship of conservation to a culture as a whole with the specific purpose of suggesting the role of environmental change, brought about by human agency, in historical geography and cultural history." He received a Guggenheim Fellowship in 1965 to study the history of the nature protection movement, an interest that was a direct outgrowth of *Traces on the Rhodian Shore.* He also edited a book for the University of Texas Press, "Reflections on the Man-Nature Theme as a Subject for Study," which was an anthology of his writings and which, for unknown reasons, was never published.[16]

Glacken earned tenure in the Department of Geography at UC Berkeley in 1964 and became chair of the department just two years later. His academic achievement, however, coincided with a period of intense political turmoil, both within academic geography and on the Berkeley campus, during a period in which he as chair was reluctantly part of the university administration. After a difficult and troubling two years of escalating faculty rivalry and student protests, Glacken had a nervous breakdown in the spring of 1970 and a physical breakdown the following fall. He took six months of sick leave. He occasionally seemed capable of a full recovery, but this proved to be chimerical. Glacken had a heart attack in 1974, after which his teaching career essentially ended. One year later Carl Sauer, who was a close friend and mentor, died, and five years later Glacken's wife, Mildred, had a serious stroke.

Despite these tragedies, his spirit was still sustained by his research and writing, and especially by his labor of love, the *Traces* sequel. While it is hard to piece together exactly what the *Traces* sequel contained, his correspondence and other fragments in his archives suggest that it had the following components:

a) A continuation of the design argument as expressed by natural theologians in the early part of the nineteenth century, including the English natural historians and geologists;
b) A study of the development of ecological ideas among the evolutionists, especially Lamarck, Darwin, and Wallace;

16. These projects are described in detail in Rajan, "Clarence Glacken."

c) A study of the subjective, emotional, and aesthetic ideas concerning the natural world in the eighteenth and nineteenth centuries, covering Keats, Shelley, Byron, and Scott, and English, French, German, and American literature;

d) An account of the environmental determinism debates, covering Henry Buckle, Friedrich Ratzel, and the French possibilists; and

e) A survey of human transformations of the environment from the nineteenth and early twentieth century to the outbreak of World War I, covering, among others, George Perkins Marsh and Nathaniel Southgate Shaler, and the early nineteenth-century German explorers of ancient Greece.[17]

What happened to the sequel is clouded in mystery. The lore in the hallways of the Berkeley Geography Department is that he did submit his final manuscript to the University of California Press in 1982 but that it was returned to him by a relatively inexperienced editor at the press with a request for substantial revision. According to this lore, which has little by way of empirical confirmation, Glacken was devastated by the rejection and destroyed all remaining copies of the manuscript. Several attempts were made to locate drafts, but repeated searches in the departmental archives turned up nothing. In 1987 Glacken moved back to Sacramento, where he remained until his death on August 20, 1989.

What most people did not know, however, is that Glacken had something else up his sleeve. He had actually written another manuscript, in addition to the *Traces* sequel. The genesis of that project can be seen in this excerpt from a letter to Frances Phillips, which he wrote around the time he was looking for a publisher for *Traces*:

> I have felt that a rather short book (about 70,000 words) written in essay style, a popular book without any sacrifice of scholarship, would be of considerable interest. It would follow along the lines of the thesis and should be of interest to laymen or scholars concerned with intellectual history and questions of population and conservation. I am aware now of many possibilities of improving my original presentation, and I know from personal experience in presenting these materials to college juniors, seniors and graduate students that they are of interest and helpful in understanding contemporary thought. This proposal would not conflict with the documented longer work which I have already described.[18]

---

17. For more details and other clues stemming from Glacken's correspondence about the sequel, see Rajan, "Clarence Glacken."

18. Clarence Glacken to Frances Phillips, Editor, William Sloane Associates, Inc., Publishers to Glacken, August 3, 1953. Glacken Papers, Bancroft Library, University of California, Berkeley, box 1, folder 1.

As it turned out, Glacken did write that book. This manuscript, written in a neat longhand, was for years tucked away in the Glacken Archives at the Bancroft Library at UC Berkeley. Titled "Man and Nature: Selected Essays," it covers twelve thinkers, from the time of the French Revolution to the mid-twentieth century. Glacken also wrote a number of papers, drafts of which he gave his colleagues David Hooson and David Stoddart. They included essays on the evolutionists, especially Darwin, Huxley, and Wallace, and a paper on Alexander von Humboldt, focusing on his subjective and aesthetic ideas. There were also essays covering the nineteenth and twentieth centuries and dealing specifically with debates relating to the habitable world, anthropogenic change, and related topics, such as carrying capacity, population, and the prospects of European migration to the New World and to other destinations in the tropics and the neotropics. Other, related topics included theories of geographic determinism, changing conceptions of soils, and the relationship between ideas of nature and human cultures.

All that remains of Glacken's unpublished works—other than the sequel, whose whereabouts remain unknown—are this handwritten manuscript and the few papers that were saved by Glacken's old friends, David Hooson and David Stoddart. These documents, referred to hereafter as the "handwritten manuscript" and the "typescript essays," constitute what might best be described as Glacken's lost works. Critically, although Glacken did not live long enough to see the emergence of the field, his writings, and these lost works, are extremely relevant to the discipline of environmental history today. That was the motivation to compile several unpublished essays in this book. The goal was to harvest the rich trove of ideas and traditions that Glacken uncovered in his quest to trace the genealogies of environmental thought.

In writing the shorter manuscript, Glacken adopted a topical or thematic approach. It was a conscious decision on his part, for he realized there were analytical trade-offs. He wrote:

> This topical procedure has its disadvantages, the most obvious being that it gives the impression that they were self-contained developments in which cross influences were at a minimum. This of course has not been the case. Population theory has been tied up with political, economic, and social theory and practice. The study of plants has had an intimate relationship to the study of soils and ecology, and it was from observations of the distribution of plants that the first modern attempts at a classification of climates were made.[19]

19. Clarence Glacken, "Introduction," in "Man and Nature: Selected Essays," unpublished manuscript, Glacken Papers, Bancroft Library, University of California, Berkeley, CU 468, 6:10. Subsequent quotations in the text are from this source.

Glacken, however, justified his method on the grounds that

> to present these in any detail would be a task of such intricacy with ideas going back and forth, changing slightly perhaps, enriched perhaps, that it would completely obscure the main outlines of my discussion in a maze of mutual influences. Even this could only be done successfully for short periods of time and even here the intricacy would be very great, since so much of the evidence would be scattered in contemporary papers and books throughout the world.

He gave two examples to illustrate such complexities. The first had to do with Malthusian theory in biology and society and with the interactions, over time, of Lyell's uniformitarianism, the research of Darwin, and then the social uptake of Malthusian ideas through interpretations of the concept of the struggle for existence in human societies—leading ultimately to ideas of *Lebensraum.* His second example was Justus von Liebig's law of the minimum, which, he pointed out, was conceived primarily in chemical terms but had been taken over into ecological theory, then expanded and modified into a concept of limiting factors, or laws of toleration, and the later work on trace elements in the soil.

Having acknowledged these methodological disadvantages, Glacken insisted that his approach could afford many tangible scholarly insights, especially by highlighting the interrelationships and the texture of the ideas that make up our concepts of the habitable world. Put simply, Glacken's method was this. He chose books, articles, and conference proceedings from the middle of the eighteenth century to about the middle of the twentieth. These sources drew on a wide variety of literature that addressed the question of the relationship of human society to the world as a whole. He admitted that these selections might or might not be representative of their time. However, he argued, they were all written by well-known authors from diverse intellectual backgrounds, ranging from philosophy to natural history, geography, ecology, economics, and sociology, and immersed in studies of global scope. Moreover, he claimed that, when complete, the book would provide "a fairly accurate picture" of the changing ideas regarding the habitable world. He reiterated this point in the concluding paragraphs of the manuscript.

This volume amalgamates the entirety of the handwritten manuscript with the typescript essays preserved by Professors Hooson and Stoddart, which follow the same method. In so doing, this book keeps to the organizational principles of Glacken's shorter manuscript, but, as mentioned in the preface, it integrates the six constituent essays into a timeline of contiguous periods. The first essay, based on the handwritten manuscript, discusses key works of three eighteenth-century writers, Charles-Irénée Castel, abbé de Saint-Pierre (1658–1743), Julien-Joseph Virey (1775–1847), and William Godwin (1756–1836). The second essay, based on the typescript essays, focuses on one towering figure, Alexander von Humboldt, and in particular on his ideas about the aesthetic and subjective appreciation of nature. The third essay, which also draws on the typescript essays, examines the ecological ideas of Charles Darwin and a few wider issues raised by some of his important contemporaries, including Thomas Huxley and Alfred Russel Wallace. Fourth is an essay based on the handwritten manuscript. It has two distinct parts. The first part discusses the importance of George Perkins Marsh in ushering in one of modern environmentalism's big themes, that of anthropogenic (human-driven) environmental changes. The second part discusses a host of other nineteenth-century writers, starting with Friedrich Ratzel, who were concerned about the carrying capacity of the Earth and possible places where Europeans could expand to and colonize. The fifth essay, drawing once again on the typescript essays, is a syncretic treatment of both these themes in the evolution of the physical and biological sciences during the nineteenth century. Last, but by no means least, the sixth essay, again based on the typescript essays, carries this discussion into the first half of the twentieth century.

For contemporary readers, Glacken's methodology might seem anachronistic. There is no attempt, in these materials, to contextualize or to situate the texts within their social and political milieus. What we have instead is a history of ideas, gleaned by way of a careful reading of key texts. Glacken was probably one of the last scholars from the tradition of the history of ideas, a tradition that became unfashionable even during the career of one of his important intellectual influences, Arthur Lovejoy.[20] Yet reading the essays in this book makes one realize that he was right in drawing attention to the persistence, through the course of more than two centuries, of themes that form the basis of modern environmentalism. Glacken does not editorialize, at least not very much. Instead, he epitomizes what a contemporary scholar has called modest witnessing.[21] This witnessing, however, is a powerful form of archeology. It uncovers parts of our past that are hidden, often in plain sight. Pondering and critically discussing the implications of Glacken's discoveries—especially those that relate to the manner in which ecological thought can be twisted to support some onerous social and political ideologies—can prove extremely profitable as humanity considers the conundrums related to the habitable world in the new millennium.

S. Ravi Rajan

20. Arthur O. Lovejoy, *The Great Chain of Being: A Study of the History of an Idea* (Cambridge, MA: Harvard University Press, 1936).
21. Donna J. Haraway, "Modest_Witness@Second_Millennium. FemaleMan©_Meets_OncoMouse™," in *Feminism and Technoscience* (New York: Routledge, 1996).

Genealogies *of* Environmentalism

PROGRESS, POPULATION, AND NATURE FROM THE AGE OF REASON TO THE MID-NINETEENTH CENTURY

This essay, based on Glacken's handwritten manuscript titled "Man and Nature: Selected Essays," consists of biographical accounts of three remarkable figures who among them spanned a two-hundred-year period, 1658 to 1847. The first, Charles-Irénée Castel, abbé de Saint-Pierre (1658–1743), was a French social reformer. He influenced a constellation of stars of the Age of Reason, including the marquis de Condorcet, Malthus, Rousseau, Voltaire, and Napoleon. He is rated one of the leading French thinkers of this period and was an advocate of the concept of the progress of scientific knowledge. He was particularly keen on deploying science to improve the human condition by harnessing the bounties of the natural world. At the same time, he cared deeply about improving both the intellectual and the cultural quality of life of humanity at large. Unsurprisingly, he was a staunch critic of despotism and superstition, and paid a political price for his views. The abbé was also strongly antiwar, and one of the earliest proponents of an international organization for world peace.

The second figure is Julien-Joseph Virey (1775–1847). Unlike the two other figures considered in this essay, who fundamentally wrote about social problems and how they might be overcome, Virey attempted to "fit human society into Nature." In his discussion, Glacken draws largely on Virey's contributions to the second edition of the *Nouveau dictionnaire d'histoire naturelle,* which was published from 1816 to 1818. Virey was a medical doctor and a professor of natural history at the Athénée of Paris when he made these contributions, which were largely about the natural world. Virey's views of nature were that it was imperfect and had strong hierarchies and conflicts. Yet there was beauty and harmony. The role of man, he wrote, was that of a peacemaker or a wise regulator. Glacken's account of Virey's writings also elaborates in great detail the key tenets of early modern environmental determinism, along with its contemporaneous purported solution, which was migration.

The third protagonist in this essay is the one best known to modern scholars, although Glacken laments that his key ideas have been lost or forgotten. In his treatment of William Godwin (1756–1836), Glacken draws almost entirely on Godwin's central polemic, *On Population*. Glacken summarizes the well-known critiques of Malthus advanced by Godwin. These included a strong methodological critique of the validity of extrapolating from one case, that of the East Coast of the United States, to the rest of the world, and many similar abuses of statistical reasoning. However, like Saint-Pierre, Godwin was a strong proponent of the idea of progress and argued that, if anything, large areas of the world were depopulated because of the inadequacies of governance stemming from a failure to deploy scientific advances. Without doubt, Godwin comes across as a tremendous technological optimist. *On Population* might well be the first of the modern classic statements on the idea of progress and development; it has many of the elements to be found in economic doctrines of development that defined state making in the twentieth century. Glacken's account of Godwin also appears to be remarkably similar to the debates between Paul Ehrlich and Julian Simon in the 1980s. Ideas might change and mutate from one context to another, but they appear forever to haunt human history.

Charles-Irénée Castel, abbé de Saint-Pierre (1658–1743)

Like so many writers who have shown an interest in ideas of this type, the abbé de Saint-Pierre lived in a turbulent age.[1] His *Annales* reflect his preoccupation with the problems of French agriculture, and they strongly condemn the wars of Louis XIV.[2] He was also concerned with the question of French depopulation and the existence of internal trade barriers, though he was not opposed to trade barriers between nations. He regarded poverty, wars, abuse of power, religious practices such as celibacy—for the Catholic Church in those times was accused of energizing depopulation—and luxury that had little social justification as obstacles to progress. To modern students, his name probably is most familiar as that of the first thinker who expanded the idea of progress to a universal principle and the author of an essay on international peace, a project to which he devoted much of his life.[3] His verbose

---

1. Charles-Irénée Castel, abbé de Saint-Pierre (1658–1743), was a French social reformer and proponent of an international organization for world peace.—Ed.

2. Charles-Irenée Castel, abbé de St. Pierre, *Annales politiques de feu monsieur Charles Irenée Castel, abbé de St. Pierre* (Paris: Londres, 1758).

3. The essay is Charles-Irenée Castel, abbé de St. Pierre, *Le Projet de paix perpétuelle* [A project for setting an everlasting peace in Europe] (1713).—Ed.

treatises on this subject were obscure until the young Jean-Jacques Rousseau undertook, in the abbé's old age, to boil down the two volumes into a shorter work, a plea for peace, to which Rousseau added some thoughts of his own.[4] The general argument that we find stated in Condorcet and Godwin, whose doctrines provoked the ideas of Malthus, were stated earlier, if anything more hopefully and more categorically, in the writings of Saint-Pierre. He had a strong influence on Rousseau, Voltaire, and, later, Napoleon.

Saint-Pierre, who is usually considered to be chronologically the first of the French thinkers of the Age of Reason, argued that since France, unlike England, did not have its revolution in the seventeenth century, ideas of progress were the only means by which despotic governments and social abuses there could be combated. What Saint-Pierre did was extend the idea of progress in knowledge to the idea of general progress in everything. The argument for the general progress of knowledge rested on two notions: (1) that the human mind was about the same in all ages, if we grant the general operation of the laws of nature, and (2) that there had been an accumulation in knowledge over the course of human history.[5] Saint-Pierre extended this concept to all fields of activity. He expected continuous and uninterrupted progress not only in knowledge but in civilization, applied science, and industry. His main thesis concerned the creation of wealth, and how that creation could go on indefinitely with good government (a sort of benevolent despotism, in the abbé's view), under rulers who followed the advice of men of science. In short, he argued that progress would follow with the removal of governmental and social obstacles that would come through knowledge and the application of knowledge in a direct and practical way. He was devoted to practical ends, as his peace proposals show. He was perhaps the first systematic utilitarian. He wrote that "the value of a book volume ... is proportional to the number and grandeur of the actual pleasures which it produces, and the future pleasures which is it calculated to provide for the greatest number of men."[6]

To understand this possibility of infinite progress, consideration of the growth and density of population was essential. He argued that within a national state, growth could be indefinite, provided that the territory was sufficient for the nourishment of the inhabitants. He claimed further that the power of a state consists

---

4. Jean-Jacques Rousseau, *A Lasting Peace through the Federation of Europe and The State of War* (1756), trans. C. E. Vaughan (London: Constable, 1917). This book actually consists of two essays on war by Rousseau. The first essay offers a critique of Saint-Pierre's idea of a European federation. The second is a theory of just war.—Ed.

5. The preceding sentences have been edited significantly for coherence.—Ed.

6. Quoted from Kingsley Martin, *French Liberal Thought in the 18th Century: A Study of Political Ideas from Bayle to Condorcet* (Boston: Little, Brown, 1929; repr., New York: Harper and Row, 1963), 61.

not in the extent of its territory but in the multitude of its inhabitants who are more laborious, more disciplined to war, more industrious in the arts, and more usefully occupied then other peoples.[7] These ideas were advanced in an age that was not as prepared for the sweeping optimisms of the latter part of the century, for there were many who still believed that modern society represented a decline from former peaks of human achievement. "He did not hesitate to predict," d'Alembert says in his eulogy of him, "that a time would arrive when, according to his own words, the simplest Capuchin would know as much as the shrewdest Jesuit."[8]

That this happy period was slow in coming did not deter him. The epochs when it could be expected to arrive could, he felt, be estimated. He undertook, as we shall see, a series of simple calculations to prove his point. The obstacles were despotism, a natural foe of knowledge and enlightenment; prejudices that the sheeplike masses adopted; and the failure of those endowed with thinking ability to heed the carnage and do so, out of a fawning respect for convention. Though a man of religion (he had entered a Jesuit order), he criticized both the Mohammedan and Christian religions in a manner that unmistakably challenged the fundamental doctrines of Christianity, with which of course he was most concerned. "To attack Mohammedanism by arguments which are equally applicable to Christianity was a device for propagating rationalism in days when it was dangerous to propagate it openly."[9] The *Annales politiques* presents the scary side of the age of Louis XIV, and his antagonism to that reign provoked the discipline of the Academy, whose members, without granting him a hearing, forbade his attendance at further meetings but without expelling him, a fate the abbé seems to have borne with equanimity.[10] His own early interest in science, and the desire to extend the scientific outlook to the government and peoples, his study of the obstacles to progress that he saw about him, his experiences as a French delegate at the beginning of the Treaty of Utrecht, his association with Bernard Le Bovier de Fontenelle, a lifelong friendship, and the conversations of the salons (in which his talk was considered as well) were the main inspirations for his modern belief in perfectibility and in the greatest good for the greatest number. "He was one of the remarkable figures of his age. We might

---

7. The source of this claim is not entirely clear.—Ed.

8. M. D'Alembert, *Select Eulogies of the Members of the French Academy with Notes, by the Late M. D'Alembert*, vol. 1 (London: Printed by A. Strahan for T. Cadell jun. and W. Davies, 1799).

9. Glacken does not provide the actual citation, although he does provide a page number. The quotation is from J. B. Bury, *The Idea of Progress: An Inquiry into Its Origin and Growth* (London: Macmillan, 1920), 77.—Ed.

10. The *Annales politiques, civiles et litteraires* was a newspaper created in 1777 by Simon-Nicolas-Henri Linguet. It was critical of the monarchy, and Linguet was punished by it.—Ed.

say that he was a new type—a nineteenth-century humanitarianism and pacifist in an eighteenth-century environment."[11]

Saint-Pierre applied such ideas to the whole globe, with optimistic conclusions. He claimed that progress is indefinite and in all fields; that population growth is desirable (and here not so subject to the limitations of a single state); that science and knowledge will convert these hopes into practical realities; that cities will grow; and that the consequent densities will produce that concentration of effort on which progress depends. The abbé de Saint-Pierre pursued these points in detail. The world, he said, could support many times its population of 900 million, whose distribution he estimated to be as follows: Asia, 360 million; Europe, 180 million; Africa, 180 million; and America, 180 million. But Europe had ten times more land than was necessary for the continent's population, Asia thirty times more, Africa fifty times more, and America one hundred times more. If we use these as multipliers (assuming that the abbé meant that if Europe had ten times more land than was needed to support its inhabitants, the population could increase ten times), the ultimate population that the world could support would reach fantastic proportions: Europe: 10 × 180 million = 1.8 billion; Asia: 30 × 360 million = 10.8 billion; Africa: 50 × 180 million = 9 billion; America 100 × 180 million = 18 billion; total = 39.6 billion. In other words, the world's population could probably increase about forty-four times over the estimate from around the 1750s. He acknowledged, however, that it would take a long time to reach this level. Wars, pestilence, earthquakes, and the like would prevent a doubling of the population in any period, and so one could not expect that the world would fill up for a very long time. And even if it did, there was equal progress in the right direction: knowledge, science, government, and morality would keep pace, and when the Earth filled up there would be ample talent to deal decisively with the situation.

Such were the ideas of a writer who lived early in the eighteenth century (he died in 1743 at the age of eighty-five). These ideas, in only slightly changed form, were those of William Godwin, who inspired Malthus, and of Condorcet, whose prestige also warranted a Malthusian rebuttal. There was no discernible Malthusianism in the abbé de Saint-Pierre. He acknowledged that wars, pestilences, and despotic governments kept the population from growing too fast, but he formulated no iron law that population constantly, in the past, present, and future, was outstripping food supply. There was little geography in his account, either: land was land, and population growth was inherently good in that it was one of the neces-

---

11. Again, Glacken does not provide the actual citation, only a number. The citation is from Bury, *The Idea of Progress*, 77.—Ed.

sary factors in supplying the density for the cities that would increase the welfare of a progressive people.

Perhaps it is the abbé's belief in an equal progression of everything that is most strange to the modern thought. Likely the increases in knowledge since his time have far exceeded his own conceptions, but it is the idea that all things go along simultaneously that has suffered the most. Those, however, who are ordained to smile at the abbé de Saint-Pierre should remember that the present age has formulated the problem in much the same way, and with the same conflicting obstacles: the march of science and the progress of knowledge, both thwarted by war, despotic government, and prejudices. If anything, a fourth obstacle has been added: the inability to reach international accord on peace or on the conservation of the world's resources.

Julien-Joseph Virey (1775–1847)

The work of the abbé de Saint-Pierre and his successors, such as the marquis de Condorcet, was concerned primarily with social questions: with the abolition of war and the possibility of knowledge and progress overcoming obstacles to progress, such as despotism and superstition. This optimistic eighteenth-century belief was unqualified in the main, and it was not until the nineteenth century that qualifications and exceptions began to appear. The previous discussion of Saint-Pierre and the following discussion of William Godwin deal almost entirely with the problems of man in society. But in the early part of the nineteenth century, attempts were made not only to discuss these questions but to fit human society into nature, so to speak, and to delineate the nature of the world in which mankind was living.

One of the best examples of this point of view, well worked out and consistent, can be found in a remarkable series of articles by Julien-Joseph Virey (1775–1847), published in the *Nouveau dictionnaire d'histoire naturelle*, the second edition of which was published from 1816 to 1818.[12] The first edition of this work was published in 1803, and the second edition was prepared to add the considerable new material that had come from advances in knowledge and discovery during the interval. It was offered to the world as a consolation and a refuge from the horrors of the war Europe had just experienced, and its editor ostensibly stated that it offered a

---

12. *Nouveau dictionnaire d'histoire naturelle, appliquée aux arts, à l'agriculture, à l'économie rurale et domestique, à la médecine, etc.* (Paris: Deterville, 1817). Julien-Joseph Virey (1775–1847) was trained as a medical doctor and served for two decades in the health service of the army; he was also a noted natural historian. Controversially, he was also a polygenist minimalist and proposed the existence of two human species, white and black.—Ed.

distraction from the horrors of war, although printing it meant overcoming several formidable obstacles.[13]

Out of this digest and study of the works of the time there came a definite and interesting picture of the nature of the habitable world. Nowhere is this seen more clearly than in Virey's articles, for, among other contributors, he was entrusted with writing those of a general nature. The writings show wide-ranging interests in eighteenth-century thinkers on social and political questions, the deep impression left by the French Revolution and the Napoleonic Wars, and the knowledge of natural exploration and its implications. When Virey wrote his articles, he was a medical doctor and a professor of natural history at the Athénée of Paris. Virey had had a military career in the Revolution, remaining in the army, where he rose to the position of chief pharmacist, until 1814. Later, he obtained his medical degree, and, after, resigning from the army (now under the Bourbon government), he taught a course in natural history at the Athénée.[14]

We are presented with a view of nature that embodies the productive principle established by divine guidance and operating under natural laws. When nature is distinguished from the works of men, we find in the latter imperfect and perishable creations that cannot reproduce themselves. Only man can understand these things, and his discoveries are to be looked upon as conquests, both glorious and innocent. The civilizing influence would come from natural history, with Linnaeus quoted approvingly. The argument was that nature occurs in a language more beautiful than that of man; that nature is permanent, whereas human things pass away; and that nature quickly conquers places usurped by man. In this world of 1816, the turbulence, fragility, and catastrophes of man brought out the consoling permanencies of nature.

Virey argued further that when we look upon organic nature, we see a hierarchy based on the laws of life and death. The destruction of beings is counterbalanced by their multiplication. Both destruction and creation are necessary. Using a line of reasoning that most writers in biology and population growth since the eighteenth century had espoused, he said that if there were no limits to natural fecundity, the universe would soon be filled up with creatures who could no longer survive because they could not destroy one another for food. Thus, there was a natural hierarchy, with very happy correspondences between this world and human society.

13. This is a paraphrasing of Glacken's original French quotation (barely legible), attributed to the volume's editor.—Ed.

14. This paragraph contains a number of scribblings in French that I have omitted because they are not legible. Based on what I could decipher, these jottings do not appear to add significant new insights to what is already reproduced here.—Ed.

The soil and the land (the terms are used interchangeably) are the matrix. The algae, lichens, and mosses that first appear are like the colonizers who prepare the land; the cereal plants are like laborers; the plants and flowers are happy citizens; the trees are like princes or kings. In this early concept of plant succession, it is difficult to see where the comparisons with the hierarchies of society stop and the interpretation of nature in terms of these hierarchies begins. For if we have colonizers, laborers, and happy citizens, and if the trees are the princes and kings of these principalities, it is easy to see what man is. He is the sovereign. It is his duty to see that the mutual destruction among plants and animals does not go too far. The plant world would be kept in check by insects, the birds, and the smaller depredators. The carnivores would exercise the next higher level of control, with man as master and sovereign.

Mankind thus stands squarely in the center of nature and in the rivalry between plants and animals, like a wise monarch over his colonists, his laborers, his citizens, and his princes of the blood. But is nature fooling herself in entrusting this great responsibility to man? Is this general hierarchy then only an expression of an eternal war of all beings against all others? Is it this tender mother that has instructed the tiger to devour the tender lamb? Does nature create these tender creatures only to submit them to an unhappy death? And what of the implications for man? For Virey, these were questions to be struggled with.[15]

However, he noted that these laws are necessary for the whole universe; they arise from the need for food, for the destruction of life to provide new life; and from the point of view of the universe this is equitable, no matter how unjust it seems in an individual case, such as the tiger and the lamb. It is this competition, this struggle in nature, that produces the overall harmony, equilibrium, and balance, under the overall guidance of a Great Being. Virey then pondered the question of why we often accuse nature of evils that are to the result of our own institutions or of a life of disorder and excess. Nature, he argued, does not work, has no desire, to create evil, but rather it is man who is endowed to ascribe to evil. Nature, if anything, is contrary to the interests of his own egotism. There is war in nature, but this is a law of life applicable to the whole universe. It results in harmony, beauty, and equilibrium, compared to the social scene. For if there is war in nature there is also mutual assistance, an order, a harmony. The disorder of nature is apparent only if one sees this concert as ineffable. This is the other side of the war in nature.

Virey went on to argue that the whole universe is a harmony in which stars constitute the members and of which we are the lesser parts. There is solidarity in this universal role of natural law, a fatal necessity. And mankind has a distinct role in this

---

15. Glacken does appear to provide Virey's answers to these questions, but they are largely illegible. Rather than reproduce sentences that do not make sense, I have omitted them.—Ed.

process. Man was created to contain the expansion of the carnivores; and the scepter has been given him to govern every being that breathes. Man is the real peacemaker in nature, and in this he is like a wise legislator who reconciles one order or society with another, creates a hierarchy of powers of mutual equilibrium that produce the calm, the harmony and happiness in the hearts of nations. In introducing mutual struggle among animals, nature has not been cruel, for has she not given guile to the weak to triumph over tyrants?[16] If we view these as harmonies, human productions are only an imitation of nature; what we call art, study, the genius and the work of man is in reality only the operation of nature, and human abilities are sharpened by the further study of nature. Real happiness comes from this redirection of oneself to nature. Tranquility is the accomplice of anyone who lives to study nature and prefers the country life to the shattering din of the cities.

With these general convictions Virey described the nature of the inhabited world as he saw it. He argued that there are the high places, sandy or hilly, that formerly were peopled but which today for the most part have been deforested and are arid or sterile. All peoples who live about these regions are nomads and act either as savages, living a pastoral existence, or as warriors or conquerors. The second class of territories in the world comprises the rich countries, crossed by fertile hills, and the valleys of light-colored earth, with their rivers and streams. These are the lands of settled society, of property rights and stable government. Such are the empires of Asia, China, Siam, Hindustan, Brazil, and so forth, whose fertility is so much that one need only scrape the surface of the soil to get abundant crops, leading to the indolence of peoples. (This idea of the fertility of the tropics and the consequent indolence of its peoples is an old, persistent, and influential one. Humboldt, as we will see, gave it considerable emphasis, and it was used by Malthus to prove his theory of land preemption.) In the interior of these empires are the highest mountains, where the soil is sterile and the peoples come more fiery and indomitable. Next, there are the territories of peoples living on the seashore littoral, in deltas, whose countries are crossed by canals and lakes—fecund peoples nourished by fish and subject to diseases of the lymphatic system. Such are the inhabitants of the shores of the Baltic, of the Low Countries, of Holland and Brabant in the middle of their polders, those at the mouth of the rivers Neman and Vistula, the lagoons of Venice, the peoples of the Nile Delta, of the Ganges, the Indus, the peoples of Mesopotamia, and those in the middle of China between the Yellow and Blue rivers. It is in Europe, according to this conception, that civilization finds its perfection because there are fewer fertile plains there than in Asia, where the soil demands only rude forms of culture. Europe is crossed by forests and mountains.

---

16. Glacken annotates here, "Is Virey writing natural history or talking politics?"—Ed.

Moreover, its peoples are less dispersed and less subdivided and therefore maintain equilibrium, a confederation that resists invasion and the establishment of permanent despotism.

Virey went on to argue that man is not controlled by climate alone: his social organization and religion contribute; however, the distributions of plants are of crucial importance, for nothing indicates better the nature of each habitation and the generalities peculiar to each country. The richness of plant life in the tropics, its rarity at the poles, produces the animal diet of the polar peoples, the vegetable diet of the tropics, the mixed diet of the intermediate zones. By special foresight, nature has sown the greater part of the nourishing grasses. The cereals and the legumes, having been distributed by nature to the temperate climate, have in turn determined the type of agriculture in these countries. It has given wheat for the cultivated fields of man, and clover and the grasses of the prairie to cattle. For this reason, the agricultural peoples are the best governed and civilized. Plants prepare the way for man, and animals obey the same laws as plants since the latter supply their food. In keeping with his claim of European exceptionalism, mentioned earlier, Virey argued that the majority of plants and animals useful to man have been placed beneath temperate skies, although the global harmonies are such that each part of the Earth offers its plants and animals a special nourishment, which in turn is tied in with the daily life and the food of men.

Virey claimed next that the place of mankind in this system is central and crucial. Man is the supreme moderator, ensuring the maintenance of equilibrium and subordination. But even when man goes too far in the increase in numbers, through the growth of civilization, nature strives to relieve herself of this burden. Virey stated: "Formerly nature was stifled in the abundance of plant and animal life; now it is overpowered, devoured by powerful hosts who exhaust the earth and its plants and destroy its animals. Then she seeks to rid herself of this exhausting multitude which oppresses her; she reverses the power of man, changes his cities into deserts by famine and disease, destroys his empires, puts, so to speak, a sword into the hands of conquerors, causes migrations from the north of destructive hoards, renews by political revolutions the mass of human generations, sends diseases which attack the reproduction of the species, and reestablishes by these formidable successes an equilibrium among organized beings."[17] In other words, nature strikes back.

We are thus presented with a world in which there is a great deal of harmony, and where harmony, once disturbed, can be restored. Man is well placed through-

---

17. Glacken does not provide an exact citation for this quotation, although he is ostensibly citing from Virey's contributions to the *Nouveau dictionnaire*. I have not independently verified the quotation.—Ed.

out the habitable Earth; the climates fit him well, even influence his temperament and civilization. The plant and animal distributions are favorable to his existence; in fact, it seems a foresight of nature that it should be so. Despite the discouragements of war and despotism, there is a great deal of hope in such a view. Catastrophes resulting from exploitation and the devastation of nature will occur, but they are parts of an ebb and flow, and we are left with the view that if it is this way in 1817, it will be this way in 2017. Man need only be like an enlightened monarch in a well-ordered society to ensure that it will go on forever. In this view, the materials are built around peoples, plants, and animals. Soils are of little importance except as the matrix for the existence of both. Population is not perpetually outstripping the food supply despite the perpetual war in nature. It is a harmonious world that we see, from the poles to the equator. It would be much later before man came to writing systemic treatises on the disastrous effects of human modifications of nature.

In Virey's system, it would seem that no irreversible changes are made by man in nature; there is no resource depletion to exhaustion. When man has gone too far, natural causes, such as famine, overpopulation, and migration, take place to restore the balance. The habitable world is thus under human supervision, and proof of this function lies in the human ability to live in all climates. But if man is despotic, he is laid low by epidemics and sudden political catastrophes. Virey argued that "Unhappy times for the human race are epochs of increase and development in the kingdoms of nature. Our multiplication and our prosperity are a period of degradation, of ruin or of decay for them, for we enrich ourselves only by depredations on nature," and as a result there is a sort of perpetual balancing, "an oscillation more or less close to equilibrium, between ourselves and the organic kingdoms." Man is over the organic world as a sovereign is over a people. But he is no different, just as sovereign and the same. Again: "It is not man who reigns on earth; it is the laws of nature of which he is both interpreter and the guardian."[18]

Virey claimed further that the human race is not perfectible, as many philosophers have thought, for the gains and losses occur at the same time; man's faculties sharpen, his body becomes enfeebled. Wise and bad customs come with the age of peoples, in this constant balancing. Moreover, this equilibrium, Virey claimed, can also be observed in the geography of the world; and changes in one part set in motion changes that affect all the others. Climate, which usually determines the temperament of peoples, determines also the spirit of each government. The peoples of the torrid zone are melancholy. Free nations are passionate, which depends on

---

18. As before, Glacken does not provide an exact citation for this quotation, although he is ostensibly citing from Virey's contributions to the *Nouveau dictionnaire*. Again, I have not independently verified the quotation.—Ed.

the nature of their territory, strengthened by a spirit of agitation and independence. Pasture peoples are more audacious, but their governments approach more to the despotic. The minority moderated by counterbalances or fixed institutions is the happy medium between democracy and despotism, and nature has especially accommodated it to the temperate climates.

This, it could be said, is the view of the world of an early nineteenth-century scientist, with faith in the laws of nature and a love of natural history—and the assurance that overall harmonies and equilibrium were the true state of nature, from the universe to the smallest living being. Virey saw in the devastation of war a reinforcement of these ideas, for there was little in nature to compare with the carnage and devastation of war. In nature there was a hierarchy and an equilibrium, as in a well-run monarchy, and man was the center of this until he too went too far and destroyed his creations. These interferences with nature Virey regarded as taking the form of the destruction of plants and animals (and to some extent deforestation); there is no hint of erosion, of the exhaustion of lands, or of upsetting the balance, except in this see-sawing between man and the organic kingdom.

William Godwin (1756–1836)

Twenty-two years had passed since the publication of Malthus's *Essay on Population* pseudonymously in 1798.[19] The wars of France and England were over. The Age of Reason seemed to have receded far into the past. The final days of the Republic and the Directory, the victories and defeats of Napoleon, were now a matter of history. The Congress of Vienna had been held. The Bourbons were in power in France. And Spain and Portugal had restored their old monarchies. The government in England had settled into a new routine of conservation and self-satisfaction. The Concert of Europe was in full swing, and throughout that continent there was that combination of political conservation and clerical influence that many historians have called the union of throne and altar. In this backwash of defeat and disillusionment, William Godwin undertook to refute Malthus once and for all. *Of Population* was published in 1820, a book of more than 600 pages that took up, argument by argument, assertion by assertion, country by country, the techniques of Malthus, which, far from lapsing into a deserved obscurity, had grown and extended their influence.[20]

Godwin began with the following declaration:

---

19. Thomas Malthus, *An Essay on the Principle Of Population* (1798, 1st ed.) *with A Summary View* (1830), *and Introduction by Professor Antony Flew* (New York: Penguin Classics, 1983). The book was published under the alias Joseph Johnson.—Ed.

20. William Godwin, *Of Population: An Enquiry concerning the Power of Increase in the Numbers of Mankind, being an Answer to Mr. Malthus's Essay on that Subject* (London: Longman, Hurst, Rees, Ornie and Brown, 1820). Subsequent quotations in the text are from this edition.

Finding, therefore, that whatever arguments have been produced against it by others, it still holds on its prosperous career, and has not long since it appeared in the impressive array of a Fifth Edition, I cannot be contented to go out of the world, without attempting to put into a permanent form what has occurred to me on the subject. (893). I was sometimes idle enough to suppose, that I had done my part, in producing the book that had given occasion to Mr. Malthus' Essay. It is stated in the front of the Essay on Population that it is to my writings that the work is indebted for its origin; and that I might safely leave the comparatively easy task, as it seemed, of demolishing the "Principle of Population," to some one of the men who have risen to maturity since the time I produced my most considerable performance. But I can refrain no longer. "I will also answer my part; I likewise will show my opinion; for I am full of matter; and the spirit within constraineth me." (v–vi)

Malthus had possessed the public mind, he said, and the old proverb, "possession is nine points of the law," was being illustrated. "*The Essay on Population* has set up a naked assertion; no more.... This author entered in a desert land, and like the first discoveries of countries, set up a symbol of occupation, and without further ceremony said, 'It's mine'" (viii). The task of refutation was difficult; it required the tactics of the siege "to dislodge the usurper," and then—referring to the celebrated lighthouse of antiquity—"to build up a Pharos that the sincere inquirer might no longer wander in the dark" (ix). This work was a practical matter, and the reader could not expect to have the "beautiful visions (if they should turn out to be visions), which enchanted my soul, and animated my pen, while writing that work" (ix) (referring to *Political Justice*).[21] This refutation, it should be understood, was not caused in a personal way. He wrote: "I hold Mr. Malthus in all due respect, at the time that I willingly plead greatly to the charge of regarding his doctrines with inexpressible abhorrence. I fully admit however the good intentions of the author of the *Essay on Population*, and cheerfully seize this occasion to testify my belief in his honorable character, and his unblemished manners" (ix).

Godwin begins his argument by appealing to the common sense and geographic knowledge of his readers.[22] Instead of Malthus discovering his population theory in the northeastern United States,[23] he argues, "Would it not have been fairer to have taken before him the globe of the earth at one view and from there to have deduced

---

21. William Godwin, *An Enquiry concerning Political Justice, and its Influence on General Virtue and Happiness*, 2 vols. (London: G. G. and J. Robinson, 1793).

22. I have removed an entire paragraph here because the text was not very clear. The take-home message, however, appears to be that Glacken is not attempting to comprehensively summarize Godwin's arguments.—Ed.

23. Malthus, Godwin said, relied on reports from North America, rather than looking for evidence from the world at large (pp. 13–14).—Ed.

the true *Principle of Population,* and the policy that ought to direct and the measures of those that govern the world?" (14). History is sacrificed for statistics. Anyone taking such a global survey would note first the thinness of human population and its immense scattering across the globe. On seeing the sorry state of the Earth as it might be, he would try to turn the uninhabited parts of the globe to better account, to see how population might be increased "and the different regions of the globe replenished with a numerous and happy race" (16). The very principle of civilized association rests on the ability to produce more than [is needed for] subsistence; if it were otherwise, we should be all cultivators of the Earth.

When Malthus ascribes the tendency of population to increase, he cannot take refuge in the law of nature, for it is not the law of nature but the law of very artificial life. As Godwin argues,

> If Malthus's doctrine is true, why is the globe not peopled, if the human species has so strong a tendency to increase, that unless the tendency was violently and calamitously counteracted, there would be everywhere "double the numbers in less than twenty-five years," and that forever, how comes it that the world is a wilderness, a wide and desolate place, where men crawl about in little herds, comfortless, unable [from] the dangers of free-loaders, and the dangers of the wild beasts to wander from climate to climate, and without that mutual support and cheerfulness which a populous earth would most naturally afford? The man on the top of St. Paul's would indeed form a conception of innumerable multitudes: but he who should survey "all kingdoms of the world," would receive a very different impression." (21)[24]

Thus was an abstract law refuted by a concrete situation. Godwin was appealing to the history of human settlement and the actual dispersion of mankind throughout the uninhabited parts of the globe. Here he has singled out a memorable distinction: for why should theories of population growth based on ideas from the experience of the United States be considered independently of the history of human populations as they have appeared and lived on the Earth's surface? This distinction between population growth (now in the hands of economists, biologists, sociologists, and demographers) and the actual distribution of population (studied by geographers and, to a lesser extent, by historians) remains with us to the present time. The core of Godwin's criticism was that Malthus was basing a natural history of population on one instance, the phenomenal growth of the American population in the latter part of the eighteenth and the early nineteenth centuries, and this theory of population Malthus presumed to hold good in the past, present, and

---

24. It appears, however, that Glacken intended here to introduce a different quotation; his annotation says the following: "(Quote here in a for Ginis illustration, Harris Lectures p -)."—Ed.

future. "The entire foundation of his work lies in one simple sentence, that in the northern states of America, 'the population there has been found to double itself, for above a century and half successfully, in less than twenty-five years'" (49).[25] It was not a question of the exception proving the rule; the exception was the rule. Certainly the rich history of the settlement of the United States and its influence on social and political thought contains fewer better domestic illustrations than this, that its early large population growth became the foundation of one of the most influential ideas of modern times, and that its settlement and cultivation throughout the nineteenth century were used to refute the validity of Malthusian theory.

One reads Godwin's *Of Population* today with mixed feelings. If he bore no personal resentment toward Malthus, this was certainly not true of his books, and the distinction between the two is a marvelous divorcement between personal and impersonal reactions. Its intolerant tone and the frequency of the declamations against Malthusian theory do not lend themselves to the dispassionate analysis that Godwin set out to perform. The book has a buckshot approach: if the duck does not fall to the ground with one salvo, it will with another. The mixture of reasoning, exhortation, biblical quotations, and peevishness makes an already repetitious book difficult to read with patience. The experience of North America was central to Godwin's refutation, yet it is hard to find what he sought to prove. "So Godwin's ideas of the American colonies vacillated between three inconsistent propositions: the great increase of the numbers is natural (or spontaneous), but that of the food is greater still; the great increase is not natural, but due to immigration; there has been no great increase at all."[26]

The idea of a geometrical increase can be tested by the experience of an Old World country such as China and a New World country such as the United States. The traditional society of China is ideal for testing a theory of population: marriage is encouraged, celibacy is discouraged, there are no manufacturing cities to produce a waste, and the quiet life of women makes them prolific and safeguards them from untimely births, according to the authority of Dr. Holde and Sir John Barrow, the latter of whom was with the Macartney Embassy in 1793.[27] China is a country where

---

25. Also relevant is the rest of the sentence, which Glacken omitted: and that this "has been repeatedly ascertained to be from procreation only."—Ed.

26. Glacken does not provide a reference to support this quotation, which he had transcribed incorrectly, but it appears in the form reproduced here in James Bonar, *Malthus and His Work* (London: Macmillan, 1885), 370.—Ed.

27. Sir John Barrow, *Travels in China: Containing Descriptions, Observations, and Comparisons, Made and Collected in the Course of a Short Residence at the Imperial Palace of Yuen-Min-Yuen, and on a Subsequent Journey Through the Country from Pekin to Canton* (London: Printed by A. Strahan for T. Cadell and W. Davies in the Strand, 1804). The book was dedicated to the Earl of Macartney, K.B.—Ed.

the true principle of population might be expected to be understood. Even though children have been exposed (left to die), the edicts against it have been strong. Godwin wonders how it is that the rulers of China, who have paid attention to the subject for centuries, "not only have no suspicion of the main principles taught in the Essay on Population but are deeply impressed with the persuasion that, without encouragement and care to prevent it, the numbers of the human species have a perpetual tendency to decline" (52).[28] The empire of China, he concludes, has never been subject to the geometrical ratio. The same is true of India, whose "institutions are gray with the hoar of many thousand years; and yet in all that time no one of her politicians and statesman has ever suspected the tremendous mischief Mr. Malthus brought to light" (55).

The attack on Malthus's authorities is largely confined to those who inspired the geometrical ratio. Benjamin Franklin, his eminence notwithstanding, cannot be considered the final authority that Englishmen would multiply like fennel.[29] Besides, Franklin was speaking out of national pride and did not fear population growth. Dr. Styles, Malthus's other authority in 1761, had merely said in a sermon that the population of Rhode Island as a whole had doubled once in twenty-five years.[30] People had been hoodwinked by the use to which the deserved reputation of the mathematician Euler had been put in the extracts from Süssmilch's work.[31] The whole principle of population was founded on one instance, with an experience lasting a bare 150 years, "in the infant colony, as I may call it in an obscure nook of the New World." "If America had never been discovered, the geometrical ratio, as applied to the multiplication of mankind would never have been known. If the British colonies had never been planted, Mr. Malthus would never have written."[32]

28. Interestingly, Glacken annotates "[Sun Yet San]" just after the quotation.—Ed.

29. Glacken does not provide a quotation here, but the following appears in Benjamin Franklin, *The Writings of Benjamin Franklin*, vol. 3, *Collected and Edited with a Life and Introduction by Albert Henry Smyth* (1750), the "Increase of Mankind," section 22: "There is, in short, no Bound to the prolific Nature of Plants or Animals, but what is made by their crowding and interfering with each other's means of Subsistence. Was the Face of the Earth vacant of other Plants, it might be gradually sowed and overspread with one Kind only; as, for Instance, with Fennel; and were it empty of other Inhabitants, it might in a few Ages be replenish'd from one Nation only; as, for Instance, with Englishmen."—Ed.

30. Glacken does not provide a quotation here, but footnote 73 in book 2, chapter 13.8, of Malthus's *Essay on Population* has the reference: "I have lately had an opportunity of seeing some extracts from the sermon of Dr. Styles, from which Dr. Price has taken these facts. Speaking of Rhode Island, Dr. Styles says that, though the period of doubling for the whole colony is 25 years, yet that it is different in different parts, and within land is 20 and 15 years."—Ed.

31. Glacken again does not provide a reference here, but book 2 of Malthus's *Essay on Population* has twenty-two references to Süssmilch on a variety of topics.—Ed.

32. Glacken does not provide references, but these quotations are from the preface to Godwin's *Of Population*. See pp. viii–x.—Ed.

Behind all this there was rebellion against the idea that population growth obeyed a natural law, that it transcended sociological theory, that doctrine lay at the base of all society, whether past, present, or future, civilized or savage. Why, he asked, are Persia, Egypt, and European and Asiatic Turkey "so thinly inhabited now, to what they were in the renowned periods of their ancient history?" (309). It was not soil exhaustion. The cause is sought in the government and administration of these countries. Many states in the past and in the Old and New Worlds are less populous now. What had the Lewis and Clark Expedition in the region of the Missouri shown? It had revealed a soil extraordinarily fertile with very few inhabitants. "Again and again captain Clark... speaks of various nations of the North American natives, the Ottoes, the Pawnees, and many more, who were once powerful races of men, but are now reduced to a feeble remnant of two or three hundred souls" (360). How are these questions to be answered? Certainly not by a law. They can only be answered with difficulty. "Population, if we consider it, historically appears to be a fitful principle, operating intermittently and by starts. This is the great mystery of the subject; and patiently to investigate the causes of its irregular progress seems to be a business highly worthy of the philosopher" (327–28).

Racial intermingling, cultural mixing, and migration also seem to have an effect on population growth. "May it not be, that races of men have a perpetual tendency to wear out? It is generally believed, both of men and animals, that a breed is materially improved by crossing, and by consequence that, where a breed is not crossed, it has a constant tendency to decline. May not the qualities of the race of Europeans such as we find them, be materially owing to the invasions of the Celts and the Cimbri, the Goths and the Vandals, the Danes, the Saxons, and the Normans?" (366).[33] And could not the decline of population in the New World be attributed to this lack of mixing, rather than to internecine war, just as historical causes such as wars and poor government accounted for areas now thinly populated that were once densely so?

There is more here than meets the eye. The world with people distributed throughout its habitable part is not to be considered the result of population pressure on food resources operating as a basic law of all life, but the habitable world as we know it is to be explained in historical and geographic terms. In their historical dispersion and their present distribution, the world's peoples have been subject to wars, invasions, and cultural intermixings, and it is these that give to one an impres-

---

33. The rest of this paragraph reads: "Perhaps, when Daniel Defoe wrote his True-Born Englishman, and thought he was composing a satire, he was very unintentionally unfolding the causes which render the natives of this island in my opinion superior, in stamina of character, in constancy of action, in intellect in humanity, and in morals, to the people of any other country now existing on Earth."—Ed.

sion of the fitfulness of population growth. This Godwin opposed to the essentially static view of Malthus, for new events, new techniques, new occurrences, good or bad, could influence population. It is a shame that these observations are buried deep in interpretations of the book, for the Godwinian approach makes possible the study of population growth not as a natural law, but as a phenomenon taking place in a part of the Earth at a definite period of time.

The final attempt to refute Malthus, Godwin clearly saw, rested in the appeal to the whole world. Malthus himself had often used the whole habitable world, especially when he wanted to dispose of arguments based on emigration. Was the world the final limiting factor? To answer this, Godwin undertook, so far as I am aware, one of the earliest modern estimates of the population capacity of the Earth, a question that every refutation of the Malthusian theory sooner or later must deal with. This Godwin did by discussing "the present state of the globe as it relates to human subsistence."[34] His point of departure is a reiteration of the previous observation, "the scanty and sparing way in which man ... is scattered over the face of the earth" (446).[35] Far from the whole surface of the Earth being like a garden, the very reverse is the case: "And it is in a world, thus cheerless and melancholy in the point of view in which we are considering it, that Mr. Malthus has thought it opportune to blow the trumpet of desolation" (447).

One should approach it in a practical way. It is estimated that the habitable parts of the world amount to 39 million square miles in area. China is said to have 1.3 million square miles of total area. China's population we can estimate to be 300 million. The extent of the cultivation is unknown; it probably could be more effectively used, but we can take "the cultivation of China for the standard of possible cultivation, and consequently its population for the standard or possible population" (449). The arithmetic is simple: $39 \div 1.3 \times 300$ million = 9 billion, about fifteen times the present population of the world. In an unpeopled world, the power of multiplication can only be viewed as a source of human happiness. In failing to see this, Malthus had really written a bitter satire of present society. It is not the Earth that limits the power of population but checks from ignorance and the institutions of society. Then there is the whole question of improving the productivity of the globe. There is no end to the improvements man can make, and it is only necessary

34. This sentence is the title of chapter 1 of book 5 of *Of Population*.—Ed.

35. The full quotation is worth reproducing here. It is: "The first thing perhaps that would arrest the observation of an enlightened enquirer, who should set himself down to survey the globe we inhabit according to the latest authorities, is the scanty and sparing way in which man, of whose nature we are, and in many respects with good reason, so proud, is scattered over the face of the earth ... When I travel even through many parts of England, it seems to me that I pass through a country, which has but just begun to be reclaimed from the tyranny of savage nature."—Ed.

to rid oneself of the geometrical ratio, to see the prospects of the cheerful world. Malthus of course grants the possibility of improving subsistence arithmetically. "But these concessions are hollow and treacherous, and the author might have known them to be so. He instantly buries his arithmetical ratio under the ponderous weight of his geometrical ratio. He might safely make these concessions; for they have had no weight with any body" (449).

More efficient land use is one answer, but the spade, not the plow, should be used. The plow would create too much leisure, and mankind is not yet improved enough to make good use of it.[36] The sea also constitutes a great reservoir of animal life for men. The grand methods on land are to substitute the plow for the pasture and the spade for the plow. "Nature has presented to us the earth, the *alma magna parens*, whose bosom, to all but the cold and incongruous ratios of Mr. Malthus, may be said to be inexhaustible" (498). To supplement this picture we must add the great advances in chemistry of the past fifty years. And then follows a statement with the same optimistic coloration that could have been made by Justus von Liebig: "No good reason can be assigned, why that which produces animal nourishment, must have previously passed through a process of animal and vegetable life" (500). Chemistry, inorganic chemistry, and human skill can ensure an indefinite improvement, leading to a world that is filled in a manner far happier than Malthus's end product. Thus it appears that, wherever earth, and water, and the other original chemical substances may be found, there human art may hereafter produce nourishment; and thus we are presented with a real infinite series of increases in the means of subsistence, to match Mr. Malthus's geometrical ratio for the multiplication of mankind.[37]

This may be thought too speculative; but surely it is not more so than Mr. Malthus's period, when the globe of the Earth, or, as he has since told us, the Solar System, and all the "other planets circling other suns," shall be overcrowded with the multitude of their human inhabitants" (500–10). Godwin's work, his biographers tell us, was unsuccessful.[38] *Of Population* is now quite forgotten, and had it

---

36. Glacken does not provide the quotation that contextualizes this sentence. The quotation, from pp. 495–96 of *Of Population*, reads as follows: "The objection I allude to is built on the consideration that, in any very improved state of human society, I should desire to see the quantity of manual labor diminished, instead of increased. But I am afraid we must pass through a probation of extensive labour, before we can come at anything better. The human species is not yet so far improved, as for the larger portion of mankind to know how to make an innocent and intelligent use of that which is, abstractly considered, the most valuable of human treasures, leisure."—Ed.

37. Glacken's sentence here is paraphrased from p. 500 of Godwin's *Of Population*.—Ed.

38. Glacken does not provide any specific reference to such critiques. However, he says: "Thomas Robert Malthus neither ignored this nor replied to it, contenting himself with a few unfaltering sentences in the last paragraph of the appendix to the sixth edition of an *Essay on Population*."—Ed.

been written later in the century, it would have been equally without influence, for it would have seemed he was flailing a dead horse. But it is a view of the world that carries over the belief in progress and adds to it the new hopes of a technological and a chemical age. There is nothing here suggesting that these advances will create any problems; there is nothing about soil, nothing about man's relationship with the world of nature. It is centered on the personality of a man who challenged the belief in the infinite progress of man. And it was a view in which the habitable Earth of 1820 was due for magnificent and cheering improvements with good institutions, with population growth, and with the final march to a real occupation of the lands of the Earth.

THE AESTHETIC AND SUBJECTIVE APPRECIATION OF NATURE
Alexander von Humboldt

This essay is based on the typewritten manuscript. It is evidently part of a sequence that includes the three thinkers discussed in the previous essay in this book, and it also refers to several others, including Count Buffon, and Chateaubriand, about whom Glacken had published essays previously, and to the English Romantics and Rousseau. According to David Hoosen, Glacken had written substantial chapters on Rousseau.[1] Glacken had also received a research grant to study the Romantic movement and its understanding of the human-nature relationship.[2]

The essay begins with some general observations about Alexander von Humboldt, including his interest in the synthesis of knowledge and in interrelationships in nature. The nub of the argument, however, revolves around the following observation: "I cannot therefore understand how historians of geography have cast Humboldt in the role of a physical scientist. That he was, but one cannot shear off, without distorting his thought, the humanistic and the aesthetic elements." Despite some other investigations, notably that of Humboldt's studies of the relationship between the histories of exploration and of science, studies Glacken claims are pioneering, the essay is largely about the humanistic aspect of Humboldt. Glacken for the most part focuses on a latter-day work, *Kosmos*, and carefully explores the treatment of landscapes there in its multitude forms, from Humboldt's own travel descriptions and evocations of the beauty and the sublime to landscape painting and the feelings it could potentially induce in the minds of people. Resonant throughout are many of the themes classically associated with the Enlightenment—self-cultivation, science, cosmopolitanism, and humanism.

---

1. David Hooson to Ravi Rajan, personal communication, November 2003.
2. Glacken to Philip Lilienthal, June 7, 1965. Glacken Papers CU 468, 1: 9.

Alexander von Humboldt's writings on nature are an indispensable part of the nineteenth-century literature on the subject, and here I wish to consider Humboldt's concept of nature as a whole. As far as I can see from his writings and from what I have read of him, Humboldt did not participate in the controversies over design in nature, nor did his ecological ideas—unlike those of Lamarck and later of Darwin—grow out of evolutionary theory. These statements do not imply that he was unacquainted with the scientific developments of the late eighteenth and early nineteenth centuries. Quite the reverse was true, as attested by his fame; his long residence in Paris, from 1804 to 1827, with some interruptions; his personal friendships with men of science; and his heavy correspondence. During Humboldt's lifetime (1769–1859), much of the thinking about nature was inspired, directly or indirectly, by the ancient idea of design. This generalization is true of Bernardin de Saint-Pierre, who believed in design, and of Lamarck, who did not.

To me, Humboldt is in that category of scientists living in the late eighteenth and early nineteenth centuries who looked forward with hope to a synthesis of knowledge of the natural world. Lamarck was one of them; so was Humboldt. One sees a common theme, strongly and persistently expressed, in the writings of both men: to view nature as a whole, to study and understand its interrelationships. If it is said that this was nothing new, that such aspirations existed as well in the ancient world and in the early modern period, I would reply that it is true that the idea of the unity of nature is ancient but that it was basically, as in the design argument, a religious and philosophical concept and belief. Lamarck and Humboldt wanted to establish this truth scientifically and to have it become the foundation of scientific research. This desire for synthesis and the knowledge of interrelationships is also related to another preoccupation of the scientists of their day, resisted by both Lamarck and Humboldt: a tendency toward the fragmentation of knowledge and, in the biological sciences, an emphasis, reflecting the continuing influence of Linnaeus, on classification.[3]

Lamarck had been quite critical of the preoccupation of biologists with taxonomy.[4] A similar attitude was apparent in Humboldt and Bonpland's *Essai sur la géo-*

---

3. Isidore Geoffroy Saint-Hilaire, *Vie, travaux et doctrine scientifique d'Étienne Geoffroy Saint-Hilaire* (Paris: P. Bertrand, 1847). On the slavish following of Linnaeus in regard to classification, see pp. 327–28. [This book is available on the Internet at www.corpusetampois.com/cse-19-isidoregsh1847viedegsh01.html.—Ed.]

4. Glacken is not claiming that Lamarck was against taxonomy, for he is known to have expanded Linnaeus's system of classification, and acquired a reputation on the basis of his work as a taxonomist. What Glacken is likely saying is that for Lamarck, there was more to biology (a term that Lamarck coined) than taxonomy.—Ed.

graphie des plantes (1807).⁵ Since I have discussed this work in *Traces on the Rhodian Shore* (pp. 543–48), I will content myself now with saying that they made a plea for a geography of plants, and that Bernardin de Saint-Pierre had anticipated them in pointing out this need. Humboldt did not disparage the botany of his time or botanists' interest in individual species, nomenclature, and taxonomy. Humboldt also wanted a geography of plants, not a geography of a single species but one of plant associations, of social plants, one that would also show altitudinal and latitudinal distributions. His famous painting of Chimoborazo (in color in the French edition) showed the plant distribution as one ascended the mountain.⁶ Humboldt did not claim that he was the first to have this idea. He acknowledged Tournefort's similar study of Mount Ararat.⁷ Humboldt's thought on the idea of a unity and harmony of nature and of interrelationships in nature is therefore closely related, although not exclusively so, to the geography of plants.

This theme of interrelationships in nature was well developed in the *Kosmos*, the work of his old age, but it was a long-held idea in his geography of plants and in his celebrated *Relation historique du voyage aux régions équinoxiales du nouveau continent, 1799–1804*.⁸ He said that he always preferred the knowledge of the interrelationships between facts observed for a long time over the knowledge of isolated facts, even if new. The discovery of a new species seemed less interesting to him than the geographic relationships of plants, the migrations of the social plants, and the altitudinal distribution of different "tribes" up to the very summit of the Cordilleras. The great problem of physical geography—*"physique du monde"*—was to determine the form of each particular type of rock, plant, or animal, the laws of their relationships, the eternal bonds that link the phenomena of life and those of inanimate nature.⁹

In this chapter, however, I wish to concentrate on the ideas of nature expressed in

---

5. The reference is Al. de Humboldt and A. Bonpland, *Essai sur la géographie des plantes, accompagné d'un tableau physique des régions équinoxiales, fondé sur les mesures exécutées, depuis le dixième degré de latitude boréale jusqu'au dixième degré de latitude australe, pendant les années 1799, 1800, 1801, 1802, et 1903 par rédigé* (Paris: Levrault, Schoelle, 1805). See the new translation by Anne Buttimer with editorial comments by Bernard Debarbieux.—Ed.

6. Humboldt, Carlos Montúfar, Aimé Bonpland, and three guides attempted to climb the Andean peak Mount Chimborazo (20,561 feet), then thought to be the highest mountain in the world, on June 23, 1802. The French edition Glacken refers to is Alexandre von Humboldt and Aimé Bonpland, *Relation historique du voyage aux régions équinoxiales du nouveau continent, fait en 1799, 1800, 1801, 1802, 1803 et 1804*, 3 vols. (Paris: Schoell, 1814–25).—Ed.

7. Joseph Pitton de Tournefort, *Relation d'un voyage du Levant: Fait par ordre du roy* (University of Lausanne, 1717).

8. Humboldt and Bonpland, *Relation historique du voyage aux régions équinoxiales du nouveau continent*.

9. Ibid., 1:33.

the *Kosmos*[10] (1845–62) because they are a distillation of Humboldt's life's thought. In the preface to the *Kosmos*, the ideas of interrelationships in nature, the role of specialized knowledge, the importance of geographic distribution, and mutual dependence are considered together. Humboldt's project was to investigate the laws relating to temperature, climate, and meteorology, and to study "the mutual dependence and connection existing between them" (1:vii). This attitude toward the study of nature was quite similar to that of Buffon, Saint-Pierre (despite his strong commitment to the design argument), and Lamarck. Humboldt characteristically saw the interrelationships in nature in philosophical, aesthetic, and scientific lights:

> In considering the study of physical phenomena, not merely in its bearings on the material wants of life, but in its general influence on the intellectual advancement of mankind, we find its noblest and most important result to be a knowledge of the chain of connection, by which all natural forces are linked together, and made mutually dependent upon each other; and it is the perception of these relations that exalts our views and ennobles our enjoyments. Such a result can, however, only be reaped as the fruit of observation and intellect, combined with the spirit of the age, in which are reflected all the varied phases of thought. (1:23)

Humboldt used the concept of a chain of being, but it is difficult to say whether it is akin to the traditional great chain of being whose history has been written by Lovejoy or whether he used the term in a looser, more general sense. What is germane to Humboldt, however, is the idea that the success of the human race in bringing a great portion of the physical world under its dominion could be looked on as a history of attempts to discover the laws of nature.

Despite their scientific character and their dependence on facts and observation, the humanistic and aesthetic aspects of nature are seldom absent from Humboldt's writings. This closeness of the scientific and the humanistic is far more than a personal predilection. A fundamental characteristic of his philosophy of nature was that there should be a meeting, not a separation, of the two. This conviction was forcefully expressed in the second volume of the *Kosmos*, in which Humboldt wrote the history of the subjective and objective views of nature. I cannot therefore understand how historians of geography have cast Humboldt in the role of a physical scientist. That he was, but one cannot shear off, without distorting his thought, the humanistic and the aesthetic elements. In this respect he resembled Lamarck: both aspired to be philosophers of nature, not only scientists.

---

10. Alexander von Humboldt, *Kosmos: A General Survey of Physical Phenomena of the Universe*, 2 vols. (London: Baillière, 1845–48). Subsequent citations are identified by page and volume number in the text.

Each of these epochs of the contemplation of the external world—the earliest dawn of thought and the advanced stage of civilization—has its own source of enjoyment. In the former, this enjoyment, in accordance with the simplicity of the primitive ages, flowed from an intuitive feeling of the order that was proclaimed by the invariable and successive reappearance of the heavenly bodies, and by the progressive development of organized beings; while in the latter, this sense of enjoyment springs from a definite knowledge of the phenomena of nature. (1:24)

Humboldt perceived that nature could be seen from different points of view. It could be considered rationally and also as a source of enjoyment. When considered rationally, nature represented "a unity in diversity of phenomena; a harmony, blending together of all created things, however dissimilar in form and attributes; one great whole animated by the breadth of life" (1:24). The enjoyment of nature had such a conspicuous role in Humboldt's thought, I believe, for two reasons: it satisfied a basic human need, and it stimulated the study of nature as an organic and physical system. In the enjoyment of nature, Humboldt shared the feelings of many pre-Romantic and Romantic thinkers. Humboldt embraced, as Wordsworth did, the idea of a communion with nature, and this was consistent with his strong conviction that no gulf should exist between the feeling for and the understanding of nature.

> In the uniform plain bounded only by a distant horizon where the lowly heather, the cistus, or waving grasses, deck the soil; on the ocean shore where the waves, softly rippling over the beach, leave a track, green with the woods of the sea; everywhere, the mind is penetrated by the same sense of the grandeur and vast expanse of nature, revealing to the soul, by a mysterious inspiration, the existence of laws that regulate the forces of the universe. (1:25)[11]

Humboldt did not introduce here the idea of a creator at all; in this omission he differed from Buffon and Lamarck, although the following quotation may have been a pro forma pronouncement, made to satisfy governmental and university authorities:

> Mere communion with nature, mere contact, with the free air, exercise a soothing yet strengthening influence on the wearied spirit, calm the storm of passion, and soften the heart when shaken by sorrow to its utmost depths. Everywhere, in every region of the globe, in every state of intellectual culture, the same sources of enjoyment are alike vouchsafed to man. The earnest and solemn thoughts awakened by a communion with

---

11. Here Glacken also references his essay on Romanticism, which is one of the lost manuscripts.—Ed.

nature intuitively arise from a presentiment of the order and harmony pervading the whole universe, and from the contrast we draw between the narrow limits of our own existence and the image of infinity revealed on every side, whether we look upward to the starry vault of heaven, scan the far-stretching plain before us, or seek to trace the dim horizon across the vast expanse of ocean. (1:25)

Such ideas were not new with Humboldt. The evocative power of nature is a theme that appears frequently in Humboldt's writing. Many passages in the *Relation historique*, often in the middle of scientific observations, describe the beauties and wonders of natural scenery in the equinoctial regions, and similar thoughts were expressed on his journey to the equinoctial regions of the New World and earlier in the *Aspects of Nature*, a popular book of travel, nature, and physical geography published in 1808.[12] Humboldt used words, such as "sublime" and "awful," that were familiar in eighteenth- and nineteenth-century descriptions of nature. He also appreciated such descriptions in the writings of others, among them Rousseau, Saint-Pierre, and Chateaubriand.

It would therefore be an error to confine oneself to an exposition of Humboldt's concept of nature based on scientific observation and experiment alone. Surely these are vital elements in this thinking because Humboldt truly believed that nature really existed and could be studied as a real entity, that it was not a construct of the human mind. One must therefore take into account, in considering Humboldt's writings as contributions to the modern literature of natural history, his subjective ideas of nature, his humanistic side, and his interest in the history of philosophy and of art. I know of no modern philosopher of nature (for this is what Humboldt fundamentally was) of whom this characterization is more true. Most writers of his time were one or the other: they were poets, prose writers, or artists, or else they were scientists. The exceptions lived in the late eighteenth and early nineteenth centuries; later, Darwin, Huxley, Wallace, and Haeckel had a scientific interest in nature; it was biological, ecological, and evolutionary, although I do not deny that aesthetic and humanistic expressions did appear in their writings, such as Darwin's description of the tropics, inspired by Humboldt, in *The Voyage of the Beagle*, and many descriptions in the writings of Wallace. Rousseau, Chateaubriand, and Wordsworth could not be regarded as scientific observations of nature, but Goethe, Buffon, and Saint-Pierre bridged the gap between the worlds of science and humanism. Goethe was deeply immersed in botany, but he also wrote *Die Italienische Reise*.[13] Many powerful literary and lyrical passages are also in Buf-

---

12. Alexander von Humboldt, *Aspects of Nature, in different lands and different climates; with scientific elucidations* (1808), trans. Mrs. Sabine (Philadelphia: Lea and Blanchard, 1850).

13. Johann Wolfgang von Goethe, with Heinrich Düntzer, *Die Italienische Reise* (Berlin, 1816–17).

fon's natural history in addition to those in *Discours sur le style*.[14] Again, despite his unrelenting commitment to the design argument, many passages in Saint-Pierre's *Études de la nature* attest to his powers of scientific observation, in addition to the artists and beauty of *Paul et Virginie*. These thinkers all lived at a time when the creative role of voyages and travels was understood and the scientific value of nature study acknowledged. These men, and Humboldt above all, saw that nature could be studied from many different viewpoints. Their attitudes were more enlightened than *Naturphilosophie*, the purely literary interpretations of nature during the Romantic movement, the utilitarian and anthropocentric preoccupations of natural theology, and the nature descriptions of skilled amateurs such as Gilbert White, whose insights were based on patient, intelligent, and acute observation.

Humboldt also had a well-developed conception of the landscape that has been prominent in nineteenth- and twentieth century geography to the present, with the renewed interest in environmental perception. These landscapes may be primitive and untouched or they may be changed by long settlement:

> The contemplation of the individual characteristics of the landscape, and of the conformation of the land in any definite region of the earth, gives rise to a different source of enjoyment, awakening impressions that are more vivid, better defined, and more congenial to certain phases of the mind, than those of which we have already spoken. At one time the heart is stirred by a sense of the grandeur of the face of nature, by the strife of the elements, or, as in Northern Asia, by the aspect of the dreary barrenness of the far-stretching steppes; at another time, softer emotions are excited by the contemplation of rich harvests wrested by the hand of man from the wild fertility of nature, or by the sight of human habitations raised beside some wild and foaming torrent. (1:25)

Elsewhere, he wrote:

> In scenes like these [described above] it is not the peaceful charm uniformly spread over the face of nature that moves the heart, but rather the peculiar physiognomy and conformation of the land, the features of the landscape, the every-varying outline of the clouds, and their blending with the horizon of the sea, whether it lies spread before us like a smooth and shining mirror, or is dimly seen through the morning mist. All that the senses can but imperfectly comprehend, all that is most awful (*das Schreckliche*) in such romantic scenes of nature, may become a source of enjoyment to man, by opening a wide field to the creative powers of his imagination. Impressions change with the varying movements of the mind, and we are led by a happy illusion to believe that we receive from the external world that with which we have ourselves invested it. (1:26)

14. Georges-Louis Leclerc, comte de Buffon, *Discours sur le style* (1753), ed. l'abbé J. Pierre (Paris: Librairie Ch. Poussielgue, 1896).

Among Humboldt's significant contributions to geographic thought, to ideas of the natural world, and to the aesthetics of landscape were his sensitive descriptions of the New World tropics. He was moved by the tropical scenes in Bernardin de Saint-Pierre's *Paul et Virginie,* although those in the *Études de la nature* are far superior.[15] Humboldt almost single-handedly made the New World tropics vivid, exciting, and scientifically fascinating places. Both Darwin and Wallace commented on the inspiration of Humboldt's writings on the tropical lands, striking illustrations of the impact of imaginative, literary, and aesthetic description of scientific thought. Humboldt himself did both: painstaking observation of the climate, geology, plants, and animals of the tropics and sparkling descriptions of what he saw and how it affected him. Of all the characteristics of the tropics, what impressed him most was the vegetation. In the following passage, one can see the mixing of scientific observation with impressionistic ideas, as the newly arrived European looked on unfamiliar landscapes.

> When far from our native country, after a long sea voyage, we set foot for the first time on a tropical land, we rejoice in the rugged wall of rock, the sight of the same kinds of mountains schistose strata, and the columnar basalt covered with cellular [amygdaloids] that we left in Europe, whose universal distribution appears to prove that the old earth's crust was formed altogether independently of the external influence of present climates. But this well-known crust is covered with an unfamiliar flora. There is revealed to us, inhabitants of the northern zone, surrounded by plants wholly unknown; the overpowering man of the tropical and exotic nature which are marvelously appropriate, energy of the human spirit. (1:24)

In the English-speaking world, awe-inspiring tropical environments are associated with Buckle's *History of Civilization in England,* but as far as I know it was Humboldt who first explored this subject.[16] The overwhelming impressions on Humboldt when seeing tropical scenery, and especially the unique characteristics of tropical vegetation, likely reinforced his theory about the evocative power of nature. Humboldt thought that certain areas on Earth were uniquely suited to evoke such impressions, and here again he broke new ground. Buffon vividly contrasted primeval nature with the well-cultivated fields of Europe, Saint-Pierre compared tropical scenery with the landscapes of Brittany and Normandy, and

15. Bernardin de Saint-Pierre, *Paul et Virginie* (Paris: Robert Laffont, 1959), and *Vœux d'un solitaire, pour servir de suite aux Études de la nature* (Paris: de l'Imprimerie de Monsieur, 1789).
16. Henry Thomas Buckle, *History of Civilization in England* (New York: D. Appleton, 1858–62). [Glacken here mentions that he will discuss Buckle in detail in a future chapter, which, alas, is also lost.—Ed.]

Captain Cook and the Forsters wrote of the dramatic environments of the South Seas. Humboldt did more: he stressed the evocative power of exotic environments. Moreover, he argued, "Graphic delineations of nature, arranged according to systematic views, are not only suited to please the imagination but may also, when properly considered, indicate the grades of the impressions of which I have spoken, from the uniformity of the sea-shore, or the barren slopes of Siberia, to the inexhaustible fertility of the torrid zone" (1:27–28, 34).The inexhaustible fertility of the torrid zone! It became a magic phrase, repeated to the present. Modern research on the nature of tropical soils has undermined this belief. Fertility seems to be derived from the cycling of energy from live to dead organic matter, and clearing the land, as in shifting cultivation, breaks the cycle.

Humboldt was at some pains to show that the sheer height of mountains was not a decisive factor in the study of nature: "But although the mountains of India greatly surpass the Cordilleras of South America by their astonishing elevation . . . they cannot, from their geographical position, present the same inexhaustible variety of phenomena by which the latter are characterized. The impression produced by the grander aspects of nature does not depend exclusively on height" (1:29). The wet tropics also have a key role in Humboldt's concept of nature. What is unique about them that makes them so vital to an understanding of the natural world? To Humboldt, the reason is that in a relatively small area on the Earth's surface, a great variety of life, especially organic life, exists; such a concentration of opportunities is found nowhere else. The wet tropics possess environments from the hot and humid lowlands to the regions of eternal snow. It is true that similar observations were made much earlier. José de Acosta in his *Historia natural y moral de las Indias* (1590) divided them into low, high, and the middle lands lying between the two extremes, corresponding to hot and humid, frigid, and temperate regions.[17] Humboldt, however, could look on these zones as a scientist supported by centuries of advances in scientific knowledge and instrumentation:

> The regions of the torrid zone not only give rise to the most powerful impressions by their organic richness and their abundant fertility, but they likewise afford the inestimable advantage of revealing to man, by the uniformity of the variations of the atmosphere and the development of vital forces, and by the contrasts of climate and vegetation exhibited at different elevations, the invariability of the laws that regulate the course of the heavenly bodies, reflected, as it were, in terrestrial phenomena. (1:34)

---

17. José de Acosta, *Historia natural y moral de las Indias: En que se tratan las cosas notables del cielo, y elementos, metals, plantas, y animales dellas, y los ritos, y gouierno, y guerras de los Indios* (Impresso en Seuilla: En casa de Iuan de Leon, 1590). See also *Traces on the Rhodian Shore*, 450.

To illustrate his point, Humboldt returned to a subject that had greatly interested him as a young man, and discussed the altitudinal zonation of vegetation in the *Essai sur la géographie des plantes*. "In the burning plains that rise but little above the level of the sea, reign the families of the banana, the cycads, and the palm," and, succeeding these, "in the Alpine valleys, and the humid and shaded clefts on the slopes of the Cordilleras," are the ferns and the cinchona. "Everywhere around, the confines of the forest are encircled by broad bands of social plants, as the delicate aralia, the thibaudia, and the myrtle-leaved Andromeda, while the Alpine rose, the magnificent befaria, weaves a purple girdle round the spiry peaks." Still higher, in the cold regions of the Páramos, is the zone of the grasses, "one vast savannah extending over the immense mountain plateau, and reflecting a yellow, almost golden tinge, to the slopes of the Cordilleras, on which graze the lama and the cattle domesticated by the European colonist." Farther up, "we meet only with plants of an inferior organization, as lichens, lecideas, and the brightly colored, dustlike *Lepraria*, scattered around in circular patches. Islets of fresh-fallen snow, varying in form and extent, arrest the last feeble traces of vegetable development, and these are followed by the region of perpetual snow, whose elevation undergoes but little change, and may be easily determined" (1:34–35).

In developing his concept of nature, Humboldt was very much concerned with the history and philosophy of science, the relationships between the humanistic and the scientific world, similar to C. P. Snow's "two worlds," and between scientific advance and the appreciation and enjoyment of nature (*Naturgenuss*). Whether because he was tactful or was unwilling to engage in polemics in his writings, he did not speak plainly about what was objectionable in the science and nature study of his day; he would have been speaking of *Naturphilosophie*, long on speculation, and perhaps even mysticism, short on painstaking, uninspiring, and tedious observation, the need for which he brought out in his descriptions of the day-to-day activities of the astronomer and the botanist. A German writer on *Naturphilosophie* has said that Humboldt rejected the purely speculative philosophy of nature that was still in vogue in his time under Schelling's and Hegel's influence for the simple reason that it forsook a basis in facts.[18] To a reader of the *Relation historique*, it would be impossible to conceive of Humboldt as a speculative thinker who systematically ignored facts.[19]

Although he was an enthusiastic admirer of general views, as were Buffon and Lamarck, he accepted and was not contemptuous of the drudgery required to

---

18. Gerhard Hennemann, *Naturphilosophie im 19. Jahrhundert* (Freiburg: Karl Alber, 1959), 107.
19. Humboldt and Bonpland, *Relation historique*.

reach a higher and more comprehensive concept of nature and its laws; he was also acutely aware of the role of inertia in thought and the persistence of ideas.

> The history of science teaches us the difficulties that have opposed the progress of this active spirit of inquiry [that is, endeavors to discover the causes of phenomena]. Inaccurate and imperfect observations have led, by false inductions, to the great number of physical views that have been perpetuated by popular prejudices among all classes of society. Thus by the side of a solid and scientific knowledge of natural phenomena there has been preserved a system of the pretended results of observation, which is so much the more difficult to shake, as it denies the validity of the facts by which it may be refuted. This empiricism, the melancholy heritage transmitted to us from former times, invariably contends for the truth of its axioms with the arrogance of a narrow-minded spirit. (1:37–38)

Humboldt contrasted this attitude with the doubts and uncertainties of physical philosophy, which "strives incessantly to perfect theory" by extending the circle of observation. These dogmas, inherited from the past, not only perpetuate error but hinder "the mind from attaining to higher views of nature" (1:38). He was, however, general in his criticisms, and it is difficult to discern exactly toward whom the shafts are being aimed—toward the special creationists, toward those who saw nature as a history of divine intercessions in natural processes at crucial times, or toward *Naturphilosophie*?

Another tendency that Humboldt thought he saw in his time and that disturbed him was the possibility of a permanent separation between science and the humanities. This was the belief that with the growth of knowledge of nature's laws, the aesthetic and emotional appreciation of nature would diminish. It would have been an exceedingly interesting episode in the history of ideas if Humboldt had been more specific and had written at greater length here as well. Did he have in mind Rousseau's well-known contempt for science and its baneful effects on one's feelings for and communion with nature, a belief that Bernardin de Saint-Pierre heartily agreed with, and did he have in mind a criticism like Lamarck's, that the circumstances of Rousseau's life had not been favorable to a just appraisal of science? Well, perhaps these queries are idle.

One cannot also neglect Humboldt's historical studies of the idea of nature; they are a key part of the *Kosmos* and, in my opinion, among the most enduring of his writings. Although I do not intend to give a complete résumé of them, I do wish to single out some of the leading concepts of the second volume of the *Kosmos* and their bearing on Humboldt's philosophy of nature. A good place to start is his history of the subjective and objective attitudes toward nature, an early, authentic classic in the history of ideas. It is a pioneer work because the history of ideas, in a technical sense, is a creation of twentieth-century thought; it was not cultivated to any degree, to my knowledge, in the eighteenth or nineteenth century. Many histories of art, philosophy, literature, and science were written, but these are not the kinds of history I mean by the history of ideas, the study of a single idea or ideas throughout a selected time span, tracing its appearance in different disciplines at different times, its relationships with other ideas, its place as a constituent part of a set of ideas.[20] It was only in 1973 that the fourth and last volume of the first edition of the *Dictionary of the History of Ideas* was published, edited by Philip P. Wiener.[21]

It is true that Darwin in *The Origin of Species* wrote a historical sketch of the ideas of his predecessors, but it was only this, not a comprehensive history of the idea of evolution. There were other tentative works, such as Alfred Biese's book on the history of the feeling for nature[22] and John T. Merz's *A History of European Thought in the Nineteenth Century*.[23] However, the self-conscious study of individual ideas is an achievement of twentieth-century scholarship. Bury's *Idea of Progress* is still a gracious classic; Teggart's *Prolegomena to History* and the *Theory of History* are as fresh as ever; Lovejoy's *Great Chain of Being* and many of his shorter essays have inspired many later efforts.[24] Humboldt's pathbreaking history therefore is all the more remarkable for its selection of a concept and following it, albeit unevenly, from early times to the first part of the nineteenth century. In no other writer in the history of Western thought concerned with nature, known to me, has the history of ideas relating to his subject been an indispensable segment of the entire philosophy.

Part of this interest in the history of ideas came from Humboldt's curiosity about the circumstances that have encouraged or discouraged certain choices in thought. Comments of this kind are frequently expressed in his writings. In the first volume of the *Kosmos* he considered the purely objective domain of the scientific delineation of nature; in the second, however, he wished to turn to a more subjec-

20. Glacken references Lovejoy here. Arthur Oncken Lovejoy (1873–1962) is often credited with having founded the discipline of the history of ideas.—Ed.

21. Philip P. Wiener, ed., *Dictionary of the History of Ideas: Studies of Selected Pivotal Ideas*, 4 vols. (New York: Charles Scribner's Sons, 1973).

22. Alfred Biese, *The Development of the Feeling for Nature in the Middle Ages and Modern Times—Primary Source Edition* (Charleston, SC: Nabu Press, 2013).

23. John Theodore Merz, *A History of European Thought in the Nineteenth Century: Scientific Thought*, 3rd ed., 2 vols. (London: William Blackwood and Sons, 1907).

24. John Bagnell Bury, *The Idea of Progress: An Inquiry into Its Origin and Growth* (New York: Macmillan, 1920); Frederick John Teggart, *Theory of History* (New Haven, CT: Yale University Press, 1925), and *Prolegomena to History: The Relation of History to Literature, Philosophy and Science* (Berkeley: University of California Press, 1916); Arthur O. Lovejoy, *The Great Chain of Being: A Study of the History of an Idea* (Cambridge, MA: Harvard University Press, 1936).

tive kind of investigation, claiming that an inner world is opened up to us (2:3). In exploring this inner world he wanted "to depict the contemplation of natural objects as a means of exciting a pure love of nature, and to investigate the causes which, especially in recent times, have, by the active medium of the imagination, so powerfully encouraged the study of nature and the predilection for distant travels" (2:19). Humboldt believed that the most effective incentives for the appreciation and scientific knowledge of nature, namely, the aesthetic treatment of natural scenery, landscape painting, and the cultivation of tropical floras, were a product of modern times, which, he believed were "characterized by general mental cultivation" (2:20, 4).

Humboldt mentioned accidental circumstances that in youth might lead one to study the natural world, referring to a passage in his *Relation historique* on the predilections one might develop at an early age for some countries and for certain climates by looking at maps and reading books of travel, and then added: "The child's pleasure in the form of countries, and of seas and lakes, as delineated in maps; the desire to behold southern stars, invisible in our hemisphere; the representation of palms and cedars of Lebanon as depicted in our illustrated Bibles, may all implant in the mind the first impulse to travel into distant countries" (2:2).[25] These objects belong to the three classes Humboldt had previously mentioned: nature description, landscape painting, and "the direct objective consideration of the characteristic features of natural forms" (2:21). These means of incitement, however, exert their power only where the state of modern culture (*der Zustand moderner Kultur*) and a unique (*eigentümlich*) course of intellectual development have, with the encouragement of original predispositions, made the feelings for impressions of nature more sensitive (2:5, 21).

In his history of ideas, Humboldt was alert to those periods that have stimulated a feeling for nature, and to differences among different cultures in their appreciation for and scientific study of nature. In his exposition of nature description among the Greeks and the Romans, Humboldt favored social rather than environmental explanations of those attitudes, even though one might expect the natural environment to inspire both the kind and the quality of nature descriptions. This expectation has a basis in fact because of the frequent idealization of the physical environment of the Greek lands, especially its climate and the interdigitation of land and sea, during the late eighteenth and early nineteenth centuries. Those who idealized the environment, however, ignored the severe Mediterranean winters and the extremely hot summers. Humboldt took issue with the limited views apparently in vogue then about the feeling for nature among the ancients. The Greeks and the

25. See also Humboldt, *Relation historique*, 1:108.

Romans were not the only peoples who had a literature; a deep feeling for nature is expressed in the oldest poetry of the Hebrews and the Indians, and therefore exists among people of very diverse, Semitic and Indo-Germanic origins (2:6–7). Humboldt showed a breadth of vision in including the ideas of other ancient peoples; the conventional approach to them—as it is perhaps even now—had been to begin with the Greeks and the Romans, and, in further European-centered fashion, to continue with the Church fathers to the late Middle Ages, and so on.

It is illustrative to read Humboldt's comment on a passage in Schiller's celebrated *Über naive und sentimentalische Dichtung*.[26] In his striking, arresting, and influential essay, Schiller had asked a pertinent question: why is it that the Greeks, living in such a beautiful natural environment, showed so little appreciation of it? The beauties of nature surrounded them. Humboldt acknowledged (in volume 2 of *Kosmos*) that when one reflects how intimately (*vertraut*) this people could live, under its happy skies, with free nature, and how much closer their mode of thought (*Vorstellungsart*), their perception (*Empfindungsweise*), and their customs were to simple nature, one must remark with astonishment that so few traces of that sentimental interest are encountered in natural scenes and the natural characteristics to which we moderns are attached (2:6).

Humboldt argued, however, that although there was much truth and excellence in Schiller's remarks, they couldn't apply (for the reasons given earlier) to all of antiquity. In saying this, he was certainly aware of the care needed in coming to conclusions based on the survival of Classical texts and the caution to be observed because of the infrequency of these expressions in their lyrical and epic poetry. His main point was that descriptions of nature really were accessories (*Beiwerk*) because in Greek art, everything moves as if it were within the human sphere (2:7). The description of nature in its rich variety of forms (*in ihrer gestaltreichen Mannigfaltigkeit*), nature poetry as a separate branch of literature was entirely foreign to the Greeks. To them, also, the landscape appeared only as the background of a painting in front of which human figures moved about. Passions, breaking out into deeds, fettered their minds almost completely. An exciting public life (*ein bewegtes öffentliches Volksleben*) drew them away from a damp, fanciful submersion in the silent forces (*Treiben*) of nature (2:25).

In reading this passage, in which Humboldt ascribes to the public life of the Greeks the failure to create a literature about nature, I am reminded of aspects of Mediterranean life, such as the promenade, that often are described in human geographies of the region, and of the contrasts often drawn between the public life

---

26. Friedrich Schiller, *Über naive und sentimentalische Dichtung* (1795) (Stuttgart: Reclam, 1957), 24–25.

of the Mediterranean peoples and the private life of those living in northwestern Europe. Aristotle's famous expression that man is a social animal (*zoon politikon*), that man is of the *polis,* has pertinence here. In any event, Humboldt called for an aesthetic appreciation of nature for its own sake, and for its evocative powers. In these passages, Humboldt reveals why he, uniquely among scientists of his time, appreciated the role of the history of ideas in formulating a concept of nature.

Wonder and puzzlement, however, remain in Humboldt's mind because it was difficult for him to understand how a people living in such an enchanting environment did not produce a literature about nature, or at least more than they did produce. Let us not forget, he said, that the Greek landscape offers the unique charm of an intimate blending of land and sea (*einer inneren Verschmelzung des Starren und Flüssigen*), "of shore adorned with vegetation, or picturesquely girt round by rocks gleaming in the light of aerial tints, and of an ocean beautiful in the play of the ever-changing brightness of its deep-toned moving waves" (2:25). The Greeks differed from other peoples, to whom the land and sea, and the occupations to which they gave rise, were two separate spheres of nature. The Greeks, not only those living on the islands but also those living in the southern part of the continent, "enjoyed, almost everywhere, the aspect of the richness and sublime grandeur imparted to the scenery by the contact and mutual influence of the two elements. How can we suppose that so intellectual and highly gifted a race should have remained insensible to the aspect of the forest-crowned cliffs on the deeply indented shores of the Mediterranean, to the silent interchange of the influences affecting the surface of the Earth, and the lower strata of the atmosphere at the recurrence of regular seasons and hours, or to the distribution of vegetable forms?" (2:25).

To these questions and puzzles Humboldt had no answers beyond the explanations he had already given. Perhaps today his perplexity would be easier to resolve. The modern study of Greek literature, advances in archeology, and the publication of discoveries on vases, metals, and other materials all reveal much more interest in the beauty and appreciation of nature than Humboldt imagined. As poetry in Greece died away along with the free life of the people, it became descriptive, didactic, a bearer of knowledge. Astronomy, geography, and hunting and fishing appear in the time of Alexander as objects of poetic art, often adorned with a very superior metrical skill. The forms and habits of animals are depicted with grace and often with such accuracy that modern natural history in classifying them can recognize genera and even species in the descriptions. All these types of poetry, however, lack the inner life, an inspired view of nature, by which the external world becomes, almost unconsciously to the excited (*angeregt*) poet, an object of the imagination (*Phantasie*) (2:12–13).

I have dwelt on this Greek illustration because it reveals Humboldt's sensitivity (rightly or wrongly) to unique conditions that inhibit the development of a nature literature and ultimately also of an objective, or scientific, attitude toward nature. Let me give one more appropriate example from the ancient world because it bears on Humboldt's sensitivity to cultural differences as they might affect attitudes toward nature. The Romans, despite their devotion to farming and the rural life (*Feldbau* and *Landbau*), were even less interested than the Greeks in the delineation of nature: "but with all their disposition to practical activity, the Romans, with the cold severity and practical understanding of their national character, were less susceptible of impressions of the senses than the Greeks, and were more devoted to every-day reality than to the idealizing poetic contemplation of nature" (2:29–30). These differences in the inner life of the Roman and Greek peoples are reflected in their literatures as in the intellectual expression of every folk-mind (2:16). Despite a relationship in the lineage (*die Verwandschaft in der Abstammung*) of the two peoples, the recognized difference in the organic structure of the two languages was associated with their cultures. Humboldt was very much impressed with the writings on language of his brother, Wilhelm, who saw in the origin of language that crucial moment when man emerged from nature. Alexander cited his work in the introduction to the *Kosmos* (1:14–15, 47). Unfortunately, as often happened in his writings on culture, Humboldt did not elaborate on this theme, which is introduced to show that cultural differences arise out of differences in language.

In his evaluations of Greek and Roman literature, Humboldt was not given to polarities. He did not deny the existence of nature writing in language either, and he commented that one language might influence the culture of another. Thus, in the Augustan period, the alienating tendency (*Hang*) to emulate Greek images inhibited the outpourings of native tenderness and a free feeling for nature; but, inspired by love of country, powerful minds knew through creative individuality, by the sublimity of their ideas, and the charm (*zarte Anmuth*) of their presentations, how to overcome these obstacles (2:16). In each period, Humboldt had his cherished authors, for no era was lacking in exemplars. Although among the Romans he recognized the merits of the nature writings of Virgil, Horace, Tibullus, and Ovid, his favorites were Lucretius and Cicero, especially Lucretius. Humboldt singled out the atomic theory and the hypothetical history of the origin and development of civilization, but it is surprising that he did not mention another outstanding characteristic of the poem, its vivid descriptions of nature, redolent of the Italian landscape.[27] The hexameral literature of the early Church fathers also impressed Humboldt, who was not noted for any partiality to exegetical theology. He had

---

27. On these points, see *Traces on the Rhodian Shore*, 29–30, 140.

praise for Basil, and justifiably so, because of the lovely descriptions of nature in Basils' writings.[28]

The age of discovery was an outstanding period in Humboldt's history of attitudes toward nature, but we need not dwell on it because what he has to say is so well known that it needs no elaboration. He compared that age with the age of Alexander, and such comparisons in Humboldt's writings usually emphasize the appearance of new ideas consequent on discoveries, their diffusion, and opportunities for comparing different and hitherto unknown environments and peoples. As one might expect from Humboldt's travels and his profound and detailed writings about them, especially in the *Relation historique,* he ranked high the discovery of the New World tropics. The tropical world, with all the luxuriance (*Ueppigkeit*) of its vegetation on the plains, with all the gradations of organic life on the slopes of the Cordilleras, with all the hints (*Anklängen*) of northern climates on the inhabited plateaus of Mexico, New Granada, and Quito, was now for the first time opened to European eyes. He was particularly touched by the simple yet vivid descriptions of nature from the pen of Columbus.

In the history of the subjective attitudes toward nature, however, Humboldt obviously considered the eighteenth century a period of efflorescence, with important preludes, particularly the development of landscape painting in the seventeenth century. Humboldt was correct in this assessment because of the preoccupation of many eighteenth-century thinkers with the natural world; it was truly a golden age of natural history, whose glory in this respect is obscured by the continuing lopsided emphasis on that century as an age of philosophy and of social and political thought. One is reminded of de Pluche, Bernardin, Pallas, Buffon, Cook, the Forsters, Banks, White, and many less well-known writers.[29]

The crucial period of the second half of the eighteenth century Humboldt described, quite naturally, more personally and intimately than previous ones; he knew Forster, Goethe, and Chateaubriand personally, and admired their original writings. He saw in the age a vigorous development of descriptive prose, and was encouraged that the accumulated masses of knowledge of nature had not had an oppressive effect on the few capable of an exalted inspiration; on the contrary, the intellectual view, the work of poetic spontaneity, had grown in scope and eminence (*Erhabenheit*). More had been learned of the structure of mountains (stratified graves of extinct organisms), the geographic distribution of animals and plants, and the relationships of peoples with one another. Among these who were able, by inciting the power of imagination, to powerfully stimulate the feeling for nature,

---

28. Here, too, see *Traces on the Rhodian Shore,* 189–95.
29. Here again, Glacken references the lost chapter on Romanticism.—Ed.

communion with nature, and the love, inseparable from it, of distant travels were Rousseau, Buffon, Saint-Pierre, and Chateaubriand in France; Playfair in Britain; and George Forster in Germany (2:65).

Humboldt's opinions of some of these men are revealing from both a scientific and a personal point of view. He was most critical of Buffon, of whom he said that:

> when he passes from the description of the habits of animals to that of the landscape, he shows in his artificially constructed periods more rhetorical pomp than individual truth to nature; rather disposing the mind generally to the reception of exalted impressions, than taking hold of it by such visible paintings of the actual life of nature, as should render her actually present to the imagination. In pursuing even his most justly celebrated efforts in this department, we are made to feel that he has never quitted middle Europe, and never actually beheld the tropical world that he engages to describe. What, however, we particularly miss in the works of this great writer is the harmonious connection of the representation of nature with the expression of awakened emotion; we miss in him almost all that flows from the mysterious analogy between the movements of the mind and the phenomena perceived by the senses. (2:64)

It is a strange judgment, and one of the very few ungenerous assessments in Humboldt's writings. He too could be pompous and could write overblown prose, if his style is judged by the standards he applied to Buffon. Some of Humboldt's sentences can be very heavy, sinking quickly to the bottom. His criticism that Buffon did not know the tropics personally, implying that this lack was incapacitating, seemingly conflicts with his later assertion that the processes of nature can be studied, learned, and understood anywhere in the world, not just in the tropics.

It is true that to a later generation, Buffon's prose may have seemed pompous, but it was also a magnificent vehicle for his ideas and is marvelously expressive. One wonders, too, how much of Buffon Humboldt had read; there are many volumes of Buffon, and one wonders whether Humboldt had read the general essays on nature, as well as the many masterful descriptions throughout the *Natural History*.[30] My surprise increased when I recalled that George Forster, whose writings Humboldt greatly admired, was himself inspired by Buffon. Forster's essay, *Ein Blick in das Ganze der Natur*, is virtually composed of paraphrases and quotations from Buffon's essay, *De la Nature: Première vue*.[31]

---

30. Georges Louis Leclerc Buffon, comte de, *Histoire naturelle, générale et particuliére* (Paris, 1749–67).

31. Georg Forster, *Ein Blick in das Ganze der Natur: Einleitung zu Anfangsgründen der Thiergeschichte* (Berlin, 1794). [Glacken refers readers to his discussion on this subject in *Traces on the Rhodian Shore*, 703. See also Clarence Glacken, "Count Buffon on Cultural Changes of the Physical Environment," *Annals of the Association of American Geographers* 50 (1960): 1–22.—Ed.]

It is difficult also for me to accept Humboldt's dictum that we miss in Buffon's writings the harmonious linking of the delineation of nature with the feelings they might awaken, that he is deficient in seeing the mysterious relationships between the emotions of the mind and the phenomena of the external world. To me, they are far more evocative than those of Rousseau, and the famous nature descriptions of Chateaubriand often are somewhat precious and mannered. Saint-Pierre is another story, for he undoubtedly had a powerful pen, and Humboldt, I think, was naturally inclined toward him because of his admiration of Saint-Pierre's skill in the delineation of the seldom described tropics.

Greater depth of feeling and a fresher life-spirit breathed in the writings of Rousseau, Saint-Pierre, and Chateaubriand. Humboldt admired the delightful eloquence, the picturesque scenes of Clarens and Meillerie on Lake Geneva in Rousseau's *Julie ou la Nouvelle Héloïse*, which, he adds gratuitously, appeared twenty years before "Buffon's *Époques de la Nature*."[32] This comment is another surprise, that Humboldt should have so cavalierly dismissed Buffon's masterpiece, a pioneer work of historical geology, containing an impressive exposition, in the seventh epoch, of human agency as a force in the transformation of nature. His most deeply felt praise, however, was for Saint-Pierre's *Paul et Virginie*, a work of a kind almost no other literature has produced, a simple description of nature on an island in the middle of a tropical sea.[33] The love story apparently did not interest him, but the accuracy and the beauty of the descriptions of tropical scenery did. In a touching passage, he says that *Paul et Virginie* accompanied Bonpland and him on their own tropical explorations, in a zone to which the novel owes its origin. For years he and Bonpland read it: "we felt ourselves penetrated by the marvelous truth with which tropical nature is described, with all its peculiarity of character, in this little work" (2:77). I have spent some time on Humboldt's opinion of Saint-Pierre because it is the best illustration, if indeed a minor one, of Humboldt's philosophy of nature, which encompassed subjective feelings, evocative power, impressions, and the scientific view of nature not as segments of unrelated knowledge but as elements of a holistic philosophy of the natural world.

A related theme is Humboldt's sensitivity to another aspect of landscape, the historical landscape; the landscape beautiful, striking, sublime in its own right but possessing the additional quality of association with great historical events (2:68). In Germany, according to Humboldt, the sole representative of these newer trends was "my famous teacher and friend," George Forster. In making this unqualified assertion, Humboldt did not ignore the German nature writers of the late eighteenth

---

32. Jean-Jacques Rousseau, *La Nouvelle Héloïse* (Paris: Le Livre de Poche, 2002).
33. Bernardin de Saint-Pierre, *Paul et Virginie* (Paris: Robert Laffont, 1959).

and early nineteenth centuries, but he found their nature poetry, in the form of pastoral idylls and didactic poems, too artificial. Stating again his conviction that one's conception of nature is derived not only from scientific observation and subjective feelings but also from the way in which these are presented by the author, Humboldt said that descriptions of nature can be sharply defined and scientifically accurate without depriving them of the invigorating breath of the imagination. The poetic must be derived from the supposed relationship of the sensuous (*des Sinnlichen*) with the intellectual; from the feeling of universality, of the reciprocal limitation and unity of the life of nature. The more sublime the objects, the more carefully must the outward embellishment of speech be avoided. The true effect of an image of nature is based on its composition, and every willful suggestion on the author's part can only be disturbing (2:74).

Humboldt was a lifelong and zealous student of the plastic arts and an accomplished artist,[34] and one of the most engaging parts of the *Kosmos* is his chapter on landscape painting and its inducements to contemplation in stimulating nature study. It is consistent with his philosophy that the aesthetic and the subjective are as important as the objective attitudes toward nature. However, although he recognized the significance of landscape painting to human thought and feeling, Humboldt discussed it in the *Kosmos* from a limited point of view: for "its role in making clear the physiognomic character of various regions of the earth" (2:76). Humboldt's botanical and meteorological investigations also strongly influenced his aesthetic views.[35]

Consistent with his characteristic desire to place subjects in historical perspective and to outline the historical circumstances favorable or unfavorable to specific developments, Humboldt commented on the status of landscape painting in the ancient world. Unfortunately, he did not expand on these topics, and his remarks are fleeting and casual. Among the Greeks and the Romans, neither landscape painting nor the poetical delineation of a region were, because of the special intellectual tendency of the two peoples, independent art forms. Subordinated to other purposes, landscape painting served for a long time as a background for historical compositions or as a fortuitous (*zufällig*) ornament for mural paintings. In a tantalizingly short passage, he sketched the history of landscape painting in Western civilization. It started with landscape as an accessory (*Beiwerk*) and gradually ending as an independent art; landscape became separated from historical painting, and the human beings served as figures (*Staffage*) in the landscape of a mountain or

---

34. Cedric Hentschel, "Zur Synthese von Literatur und Naturwissenschaft bei Alexander von Humboldt," in *Alexander von Humboldt: Werk und Weltgeltung*, herausgegeben von Heinrich Pfeiffer für die Alexander von Humboldt-Stiftung (München: R. Piper, 1969), 31–95, at 73.

35. Hentschel, "Zur Zynthese," 75.

forest region, a strand of the sea or a garden park. From these summary generalizations, Humboldt concludes that the feeling for the beauty of landscape (*das Gefühl fur die landschaftliche Schönheit*), which the brush restores to us, is not an ancient but a modern feeling (2:77). The effect of this sharp distinction, according to Humboldt, was to claim a break, not a continuity, between ancient and modern feelings for nature.[36] He made another significant distinction between Greek and Roman times and modern times. Only what was agreeably habitable in the landscape was attractive to the Greeks and Romans, not what we call wild and romantic (2:79). If we bear in mind the crucial importance that Humboldt attached to the eighteenth-century nature writings of Rousseau, Buffon, Saint-Pierre, and Chateaubriand and add to them the efflorescence of landscape painting in the seventeenth century, we have a heady and creative mixture that distinguishes these centuries, in the history of Western civilization, both in the scientific study of, and in the subjective feeling for, the world of nature. Humboldt's philosophy of nature called for purposive activity as a way of continuously stimulating feelings for the beauties and wonders of nature. To him, panoramas were admirably suited for such needs. Panoramas, hitherto used more frequently to depict cities and inhabited areas, could well be used with enchanting effect for rural scenes as well, and "for the rugged declivities" of the Himalayas and the Cordilleras (2:98).

In concluding his essay, Humboldt characteristically relates landscape painting not to science or to an elite but to all people. It was as if he realized that an understanding of and a feeling for nature are meaningless unless they are widely diffused and made an integral part of civilization. His summary reveals satisfaction in the knowledge of the natural world, subjective and objective, that had been achieved in human history, and especially in the seventeenth and eighteenth centuries. The knowledge and the feeling, he says, for the sublime grandeur of the creation (*vor der erhabenen Grösse der Schöpfung*) would be powerfully increased if in our great cities, besides the museums, but like them open free to the public, a number of panoramic buildings (*Rundgebäuden*) exhibiting alternating landscapes and from different latitudes and altitudinal zones were erected. The concept of nature as a whole (*Naturganzen*), the feeling of unity and of harmonious accord throughout the cosmos, "cannot fail to increase in vividness among men, in proportion as the means are multiplied by which the phenomena of nature may be more characteristically and visibly manifested" (2:91, 98).

Despite his own intense interest in landscape painting and its historical rela-

---

36. Glacken argues here that this point has also been made by art historians who were his contemporaries, such as Kenneth Clark, in *Landscape into Art* (London: John Murray, 1949), 141.—Ed.

tionship to the stimulation of feelings for nature among human beings, Humboldt thought that it was less stimulating in evoking feelings for nature than were exhibitions of exotic plants in hothouses and in gardens, and he recalled how the sight of a dragon tree (*Drachenbaum*) and a fan palm in an old tower of the Botanical Garden in Berlin had implanted in his mind the first seed of an irresistible longing for journeys to far-off places (2:95). Humboldt's comparisons of the impressions received from landscape painting and from exhibitions of plants are models of sensitive perception, and one can only wish that he had elaborated more on these themes. He wrote that landscape painting can present a richer, more complete image of nature than can the artificial groupings of cultivated plants. However, although the multiplication of means at the disposal of painting to incite the imagination and to concentrate, as in a small space, the great phenomena of sea and land is denied to our plantations and garden parks, they compensate singly through the mastery (*Herrschaft*) that reality everywhere exercises over the senses, even if the total impression of the landscape is less (2:96–97).

In theorizing as to the origin of parks in Central and South Asia, Humboldt pointed to the historical importance of the worship of trees. Because of the refreshing and humid shadow of the leafy canopy, the ancient worship of trees was associated with the worship of sacred springs. He noted the fame and veneration of the Indian fig tree, the remarkably large palm tree in Delos, another in Arcadia, the colossal fig tree of Anuradhapura, Sri Lanka, and the worship not only of single trees but of sacred groves.[37] These are themes that have played significant roles in human geography of the twentieth century. Humboldt was also interested in Chinese culture. He had not yielded, he said, to the distaste for Chinese literature prevalent among his contemporaries, and Chinese gardens especially attracted him. In the early and exquisite cultivation of the Chinese garden he saw a manifestation of a universal human need.[38] Here he remarked that through the centuries, people have agreed that the plantings (*Pflanzung*) of men should compensate for the loss of all those charms that man's distancing (*Entfernung*) from the life of free nature, his own and most lovely abode, has deprived him of (2:100). Anticipating a theme in modern religious geography, Humboldt mentioned the role of Buddhism and the intimate interrelationships existing among Buddhist monasteries in the culti-

---

37. The Indian fig tree is the banyan (*Ficus benghalensis*), the species Hindus consider sacred. The tree in Anuradhapura is a bodhi or peepal tree (*Ficus religiosa*) and is said to have grown from a cutting of the tree in Bodh Gaya, India, under which Siddhartha Gautama ostensibly attained enlightenment. The large palm tree in Delos is located by the sacred lake near which Leto is said to have given birth to the Greek god Apollo and his twin sister, the Greek goddess Artemis.—Ed.

38. Glacken notes that Humboldt drew on secondary sources for his writings on China, and references Humboldt, *Kosmos*, 2:133.—Ed.

vation and distribution of plants and in the laying out of gardens. Temples, cloisters, and cemeteries were surrounded by garden parks, adorned with exotic trees and a carpet of multicolored and multiform flowers. Indian plants diffused early to China, Korea, and Japan (2:102–3, 134).

Humboldt, however, has nothing to say of the European garden; of the French vista garden, such as Versailles; or of the English garden. Nor is there any effective exposition of the relationship of gardens to nature and society.[39] However, Humboldt saw in the European culture of his time circumstances favorable to furthering an understanding of nature in all of its aspects:

> We may therefore regard it as one of the most precious fruits of European civilization, that it is almost everywhere permitted to man, by the cultivation and arrangement of exotic plants, by the charm of landscape painting, and by the inspired power of language, to procure a substitute for familiar scenes during the period of absence, or to receive a portion of that enjoyment from nature which is yielded by actual contemplation during long and not infrequently dangerous journeys through the interior of distant continents. (2:105)

To Humboldt, this achievement clearly was an important part of the civilizing process.

The *History of the Physical Worldview* (*Geschichte der physischen Weltanschauung*) is slightly more than twice as long as the preceding *Incitements to the Study of Nature* (*Anregungsmittel zum Naturstudium*)—266 pages in the former case, 132 in the latter. Despite many excellent passages in the second essay, to me the first remains fresher and more creative; even today its subject matter will be new to many, while that of the second (through no fault of Humboldt) is more traditional in its elements and far better known. Nevertheless, the essay on the physical worldview possesses a novelty and an originality in presentation that make it distinctive in the history of historiography. It is really the history of two interlocking histories, the

---

39. Glacken notes here that the garden as a manifestation of a philosophy of nature is a recent field of research, and that although many writers in the eighteenth and nineteenth centuries recognized in the garden expressions of attitudes toward the natural world, its historiography, and especially its relation to philosophical ideas, are twentieth-century achievements. He mentions the works of Arthur Lovejoy and Derek Clifford as examples of how the garden can be a subject for the history of ideas. He is likely referring to Arthur O. Lovejoy, "The Chinese Origin of a Romanticism," *Journal of English and Germanic Philology* 32, no. 1 (1933), and Derek Clifford, *A History of Garden Design* (New York: Praeger, 1963). Glacken also mentions the genre of national histories of gardens, and cites the works of Georgina Masson, Edward Hyams, and Elizabeth Manwaring. Here he is likely referring to Georgina Masson, *Italian Gardens* (London: Thames and Hudson, 1961; repr., Woodbridge: Garden Art Press, 2011); Edward Hyams, *The English Garden*. (London: Thames and Hudson, 1964); and Elizabeth Manwaring, *Italian Landscape in Eighteenth Century England* (London: Cass; New York: Russell and Russell, 1965).—Ed.

history of exploration and the history of science. In Humboldt's presentation, they become so interdependent that one cannot be understood without the other. The intermeshing of these two histories, beginning with the Mediterranean world and continuing to his own time, was a remarkable methodological innovation. Such histories are not even written today; the history of science and the history of exploration are separate and distinct fields of research. I am not saying that histories of science are composed without reference to the history of exploration, or that histories of exploration are concerned with the discovery, occupation, and description of new lands to the exclusion of their influence on theoretical science, but that Humboldt's handling of the subject goes far beyond expected and predictable overlappings; both are on an equal footing.

Although I have no passage to point to as proof, I suspect Humboldt's reason for joining together the history of exploration and the history of science was his fundamental concern with ideas and their diffusion, and here the geography comes in, the influences of the configuration of the Earth's surface, the presence of bodies of fresh- and salt water, and the like on the routes of this diffusion. Closely related to this interpretation was Humboldt's conviction, inspired by his brother, Wilhelm, of the crucial role of language as a vehicle not only for the expression of ideas but also for their diffusion. Although Humboldt was not an exponent of historicism, the conviction is conspicuous in his writings that history is a powerful tool not only for explaining social, political, geographic, and physical conditions but also for understanding thought and ideas (2:106).[40] He was also impressed by the diffusion of ideas, cultural contact, and the increased knowledge of cultural differences consequent on the campaigns of Alexander the Great and the Hellenistic period that followed. Speaking of the influence of these campaigns, he said:

> The enlargement of the sphere of ideas [*Ideenkreis*], which arose from the contemplation of numerous hitherto unobserved physical phenomena, and from a contact with different races, and an acquaintance with their contrasted forms of government, was not, unfortunately, accompanied by the fruits of ethnological comparative philology, as far as the latter is of a philosophical nature depending on the fundamental relations of thought, or is simply historical. This species of inquiry was wholly unknown to classical antiquity. (2:166)

Humboldt cited contemporary scholars on comparative philology, especially of the Indo-Germanic languages (2:431n6).

Humboldt never conceived of thought as something disembodied, existing

---

40. See the footnote about historicism on this page.—Ed.

by itself, remote from observed reality; he revealed himself as a person close to is the observed world and sensitive to the power of instruments such as the compound microscope, the barometer, and the telescope, and to the imaginative and creative role of subjective impression. He also fundamentally believed in the interdependence of knowledge, especially though contact between human peoples through history, and that primitive peoples may originally have had knowledge of the physical constitution of the universe: "True cosmical views are the result of observation and ideal combination, and of a long-continued communion with the external world; nor are they a work of a single people, but the fruits yielded by reciprocal communication, and by a great, if not general intercourse between different nations" (2:116). Humboldt postulated that the incitements to the study of nature, the subjective impression, led to curiosity about natural processes, which then led to a search for scientific laws and finally to the objective study of nature. It in turn further stimulated subjective feelings, the result being a continuing mutual reinforcement of the two ways of perceiving the natural world. This conception seems related to the idea of feedback (2:117).

In his history of the physical view of the universe, Humboldt chose, from the bewildering mass of historical events, seven principal moments (*Hauptmomente*), or epochs. The scheme was European-centered because he regarded Europe as the permanent home of civilization, a point of view easier to maintain in the middle of the nineteenth century than in the middle of the twentieth. These seven moments were:

1. The Mediterranean Sea as the point of departure for the exposition of relationships that have laid the foundation for the gradual extension of the idea of the cosmos;
2. The Macedonian campaigns under Alexander the Great;
3. Advances in worldview under the Ptolemies, a moment that would now be called the Hellenistic period, so named by Johann Gustav Droysen, the German historian and a contemporary of Humboldt, whom he cited in this chapter;
4. Roman world hegemony;
5. The Arab invasion;
6. The era of the oceanic discoveries; and
7. Great discoveries in space through the use of the telescope.

I do not intend to summarize Humboldt's detailed discussion of each of these periods. The *Hauptmomente*, however, are an interesting example of periodization in history. Of the seven moments, the first was based on geographic position, the second through the sixth on the spread of ideas owing to cultural contact, and the last on astronomical discovery and instrumentation.[41] To Humboldt, the age of discovery brought the first six epochs to a close but "terminated with the acquisition of an entire hemisphere which had till then [had] lain concealed, and which constituted the greatest geographic discovery ever made" (2:353). I wonder why Humboldt made such a sharp break. Large parts of Asia, Africa, the Americas, the South Seas, and the polar lands remained to be explored scientifically, to say nothing of the ethnological research that had barely got under way and that flourished dramatically only in the decades after Humboldt's death.

As one might expect from his sustained interest in the subject, the longest chapter in Humboldt's essay on the history of the physical worldview is on the oceanic discoveries. Of particular interest to me is his sensitivity to the wide range of their effects, and especially to the importance of cultural contact and the consequent diffusion of ideas. The age of discovery doubled the number of works of creation known to Europeans; it offered new and powerful incitements to the study of mathematics and physics; "terrestrial nature was conceived in its general character, and made an object of direct observation;" regions and constellations in the heavens were discovered; the establishment of comparative geography was now possible; "extraordinary changes in the manners and well-being of men" were brought about, including the late awakening of enslaved peoples to the political; the field of view expanded and extended, owing to trade and colonization (2: 228–29).

Did Humboldt believe in the idea of progress? He was optimistic about the future and about the prospects for an increase in knowledge, but he did not seem to have a well-defined concept of progress. It was closely related to his conviction that the study of nature, cosmic and terrestrial, was open-ended; the nineteenth century may have been a climax, but it was no end:

> Excited by the brilliant manifestation of new discoveries, and nourishing hopes, the fallacy of which often continues long undetected, each age dreams that it has approximated closely to the culminating point of the recognition and comprehension of nature. I doubt whether, on serious reflection, such a belief will tend to heighten the enjoyment of the present. A more animating conviction, and one more consonant with the great destiny of our race, is that the conquests already achieved constitute only a very inconsiderable portion of those to which free humanity will attain in future ages by the progress of mental activity and general cultivation. (2:354–55)

41. On this subject, Dietrich Gerhard, "Periodization in History," in Wiener, ed., *Dictionary of the History of Ideas*, 3:476–81; and Rene Wellek, "Periodization in Literary History," ibid., 3:481–86, although Humboldt is not mentioned.

The humanism of Humboldt—his love of freedom, his hatred of slavery, his association of the growth of science with an increasingly sensitive awareness of nature, his cosmopolitanism—made his history of ideas possible (2:355). Humboldt had higher goals than achieving an objective knowledge of the natural world. Those goals are apparent throughout his writings. Without being narrowly anthropocentric, his views of objective nature were closely tied in to the inner life, to subjective völker feelings—the fusion of the two being of a higher order than either one singly. Ernst Haeckel saw this quality in Humboldt. In both the *Kosmos* and the classic *Ansichten der Natur*, he said, Humboldt had joined together in the happiest fashion the scientific and aesthetic views of nature, and had correctly emphasized how closely the exalted enjoyment of nature is linked to the scientific research on cosmic laws and how both, joined together, served to elevate the human condition (*Menschenwegen*) to a higher stage of perfection: "The astonishing wonderment with which we observe the starry heaven and microscopic life in a drop of water, the awe with which we investigate the wonderful working of energy in active matter (*der bewegten Materie*), the devotion with which we revere the validity of all embracing laws of matter in the universe—they all are essential parts of our internal life included in the concept of 'natural religion.'"[42] In viewing the cosmos as a whole, Humboldt welcomed progress in mathematics. To Humboldt, who was no theoretician, mathematically formulated hypothesis and theories represented the highest stage in the natural sciences.[43] As far as living nature on Earth was concerned, he highly valued the concrete, the specific. The basic elements of the "comparative physiognomy" of plants, or what we today call ecology, were, however, based on detailed field research.[44]

In his correspondence with friends, above all with Goethe and Varnhagen von Ense, Humboldt repeatedly alluded to the aesthetic elements in his writings; when he sent Goethe a copy of his *Pittoreske Ansichten der Cordilleren und Monumenta amerikanischer Völker*, he emphasized that man must feel nature, to be delivered from bloodless abstraction.[45] To me, the best and most dramatic illustration of Humboldt's philosophy of nature is not to be found in the *Kosmos* at all but in the

42. Ernst Haeckel, *Gemeinverständliche Werke*, 6 vols. (Leipzig: Alfred Kröner Verlag; Berlin: Carl Henschel Verlag, 1924), 3:355.
43. Mario Bunge, "Alexander von Humboldt und die Philosophie," in *Alexander Humboldt: Werk und Weltgeltung*, 27.
44. Adolf Meyer-Abich, "Alexander von Humboldt as a Biologist," in *Alexander von Humboldt: Werk und Weltgeltung*, 179–96; on comparative physiognomy, 190.
45. Alexander von Humboldt, *Pittoreske Ansichten der Cordilleren und Monumenta amerikanischer Völker* (Tübingen: Cotta, 1810).) [Glacken suggests that the date of the letter to Goethe was January 3, 1810, but does not provide a credible source. The sentiments expressed here are, however, consistent with what is known about Humboldt and Goethe. See, e.g., Laura Dassow Walls.

Ansichten der Natur, an early, popular series of essays, which was his own favorite. It is a passage from "The Nocturnal Life of Animals in the Primeval Forests" (Das nächtliche Tierleben im Urwalde).[46] For twenty years I read it to my classes in the history of ideas of culture and nature. Humboldt, Bonpland, and their party had passed the night in the open air below the mission of Santa Barbara de Arichuna on the Apure river, "skirted by the impenetrable forest," in present-day Venezuela. After they had settled down,

> a deep stillness prevailed, only broken at intervals by the blowing of the freshwater dolphins peculiar to the river network of the Orinoco. After eleven o'clock, such a noise began in the contiguous forest that for the remainder of the night, all sleep was impossible. The wild cries of animals rang through the woods. Among the many voices that resounded together, the Indians could only recognize those which, after short pauses, were heard singly. There was the monotonous, plaintive cry of the Aluates (howling monkeys), the whining, flutelike notes of the small sapajous, the grunting murmur of the striped nocturnal ape (*Nyctipithecus trivirgatus,* which I was the first to describe), the fitful roar of the great tiger, the cougar or maneless American lion, the peccary, the sloth, and a host of parrots, parraquas (*Ortalides*), and other pheasant-like birds. Whenever the tigers approached the edge of the forest, our dog, which before had barked incessantly, came howling to seek protection under the hammocks. Sometimes the cry of the tiger resounded from the branches of a tree, and it was then always accompanied by the plaintive piping tones of the apes, which were endeavoring to escape from the unwonted pursuit. If one asks the Indians why such a continuous noise is heard on certain nights, they answer, with a smile, that "the animals are rejoicing in the beautiful moonlight, and celebrating the return of the full moon. To me the scene appeared to owe rather to an accidental, long continued, and gradually increasing conflict among the animals. Thus, for instance, the jaguar will pursue the peccaries and the tapirs, which, densely crowded together, burst through the barrier of treelike shrubs that oppose their flight. Terrified at the confusion, the monkeys on the tops of the trees join their cries with those of the larger animals. This arouses the tribes of birds that build their nests in communities, and suddenly the whole animal world is in a state of commotion. Further experience taught us that it was by no means always the festival of moonlight that disturbed the stillness of the forest, for we observed that the voices were loudest during violent storms of rain, or when the thunder echoed and

---

"Introducing Humboldt's Cosmos," *Minding Nature* 2, no. 2 (August 2009): 3–15. Karl August Varnhagen von Ense (1785–1858) was a writer and biographer of renown.—Ed.]

46. Alexander von Humboldt, *Ansichten der Natur mit wissenschaftlichen Erläuterungen* (Stuttgart: Cotta, 1849), chap. 10.

the lightning flashed through the depths of the woods.... A singular contrast to the scenes I have here described, and which I had repeated opportunities of witnessing, is presented by the stillness that reigns in the tropics at noon of an unusually sultry day.[47]

What follows is the description of a scene at the narrows of Baraguan:

A thermometer observed in the shade, but brought within a few inches of the lofty mass of granite rock, rose to more than 122° Fahr. [mehr als 40° Reaumur]. All distant objects had wavy undulating outlines, the optical effect of the *mirage*. Not a breath of air moved the dustlike sand. The sun stood at the zenith, and the effulgence of light pouring upon the river, and which, owing to a gentle ripple of the waters, was brilliantly reflected, gave additional distinctness to the red haze that veiled the distance. All the rocky mounds and naked boulders were covered with large, thick-scaled iguanas, gecko-lizards, and spotted salamanders. Motionless, with uplifted heads and widely extended mouths, they seemed to inhale the heated air with ecstasy. The larger animals at such times take refuge in the deep recesses of the forest; the birds nestle beneath the foliage of the trees, or in the clefts of the rocks; but if in this apparent stillness of nature we listen closely for the faintest tones, we detect a dull, muffled sound, a buzzing and humming of insects close to the earth, in the lower strata of the atmosphere. Everything proclaims a world of active organic forces. In every shrub, in the cracked bark of trees, in the perforated ground inhabited by hymenopterous insects, life is everywhere audibly manifest. It is one of the many voices of nature revealed to the pious and susceptible spirit of man.[48]

A history of the subsequent fortunes of Alexander von Humboldt's writings would quickly evolve into an arresting commentary on the history of science, aesthetics, and humanistic thought. After his death, his fame and influence declined, so much so that by the end of the century his name was little remembered. Part of the decline was inevitable: much of the science in the *Kosmos* soon became obsolete; he died in the same year, 1859, in which *The Origin of Species* was published; and the history of science in the last half of the nineteenth century is to a remarkable degree a history of controversy about, comment on, supplements to, and revisions of evolutionary theory. There was no hint in Humboldt's writings, so far as I can determine, of a theory of evolution in the sense of Lamarck or Darwin, but there was a sympathy with an older idea of development prominent in the eighteenth and early nineteenth centuries among thinkers such as Herder. Humboldt was an admirer of Malthus's work on population; he mentioned it approvingly in the *Rela-*

47. Ibid., 199–201.
48. Ibid.

tion historique and in the *Political Essay on New Spain*.⁴⁹ Unlike Darwin and Wallace, however, Humboldt did not extend the Malthusian theory to the biological world.

Humboldt was not completely forgotten. Even if his brother's works, and not his, survived in histories of German literature, he remained a hero in Latin America, and his name and accomplishments endured in Latin American studies, in university curricula in geography, and in the history of geography. The centenary of his death in 1959 saw an impressive revival of interest in Humboldt, inspired almost entirely by German scholars and scientists. A new era of Humboldt scholarship began, the general tenor of which has been to stress the manifold character of Humboldt's contributions to, among other disciplines, philosophy, plant geography, ecology, literature and humanistic studies, history, and ethnography.⁵⁰

---

49. Alexander von Humboldt. *Political Essay on the Kingdom of New Spain: With Physical Sections and Maps Founded on Astronomical Observations and Trigonometrical and Barometrical Measurements* (London: Longman, Hurst, Rees, Orme, and Brown, 1822), 1:197.

50. Glacken's manuscript has rough notes from some of these sources, and a fairly long description of his philosophy of science, summarizing from Bunge, "Alexander von Humboldt und die Philosophie."—Ed.

DARWIN AND HIS CONTEMPORARIES

This essay is an amalgamation of two typewritten essays, one on Charles Darwin and another on Darwin, Huxley, and Wallace. It is organized in three parts. In the first, Glacken examines Darwin's philosophy of nature. Darwin's critique of the design argument, a theme that continues from *Traces on the Rhodian Shore*, constitutes one element of the discussion, culminating in a discussion of Darwin's "biological *Candide*," a phrase Glacken borrows from the philosopher Michael T. Ghiselin. In this section Glacken also explores Darwin's use of metaphors, and discusses the place of metaphors in science more generally. Yet another theme is the tension between the natural and the social. Here, interestingly, and probably because of his method—a close textual study of Darwin's writings, including his published letters—Glacken categorically rejects the suggestion that nineteenth-century social thought had a role to play in Darwin's idea of nature.

The second section of the essay examines Darwin's ecological thought, concentrating on chapter 3 of *The Origin of Species*. Arguing that the concept of a struggle for existence was in essence similar to an older term, the economy of nature, Glacken focuses here on some of the principles of ecology enunciated in this essay. These include key themes that are resonant today, including the concept of adaptation. This section also examines the various influences climate has on ecology, the concept of geographic distribution, and the groundwork for evolutionary theory laid by Darwin. Last but by no means least, this section discusses Darwin's aesthetics, and his appreciation of the beauty of plant and animal life.

The third and concluding section of the essay explores the legacy of Darwin and that of two of his contemporaries, Huxley and Wallace. It explores the concept of the balance of nature, and many related themes in ecological theory, before broadening the discussion. Among the topics addressed are the reemergence of Malthusian ideas in Huxley's thought, the three men's thoughts on the differences between humans and apes, and the difference between moderns and primitive peoples. Glacken attempts here to differentiate between these otherwise kindred spirits, and approvingly mentions Wallace's (and to a degree Huxley's) sensitivity to the conditions of the working classes.

Darwin's Philosophy of Nature

Although I realize that Darwin's thought is inseparable from his theory of evolution, in this essay I wish to place his philosophy of nature, which was based on concepts of interrelationships in the natural world, a manner of thinking now called ecological, at the center, as I did in my essays on Saint-Pierre, Alexander von Humboldt, and others.[1] My argument is that although an ecological point of view can be discerned in the physico-theology of the seventeenth and eighteenth centuries, its modern scientific basis, founded on a theory of evolution and not on the special creation of design, is derived from Darwin. Buffon, Lamarck, and Humboldt were important precursors who, like Darwin, were opposed to special creation and teleology in nature, but their contributions were overwhelmed by the ultimate success and dominance of Darwinism. My observations here are based on his well-known works, and I doubt whether further research or the discovery of new sources will seriously alter his ideas regarding ecology, or, to use the popular phrase of the time, the economy of nature, expressed in already published writings.

I cannot refrain from stating my own attitude toward Darwin at the very outset. The more times I read *The Origin of Species,* the more impressed I am with the wealth of observation and thought on which his argument, despite obvious lacunae, is based, and with his adroit use of the evidence.[2] *The Descent of Man* reveals a wide and through study of writing of the leading European prehistorians and anthropologists of his day.[3] *The Variation of Animals and Plants under Domestication* is an astonishing work that, along with George Perkins Marsh's *Man and Nature* (1864), is one of the most impressive contributions to the literature on the modification of nature (in this case, the domestication of animals and plants through human agency that

---

1. Glacken also alludes to an essay on Lamarck, now lost.—Ed.

2. Charles Darwin, *On the Origin of Species by Means of Natural Selection* (London: John Murray, 1859). Subsequent citations in the text are to this edition by page number. [Here Glacken, in a self-referential note, comments, "This quality will be apparent to anyone who himself has composed a work of synthesis whose sources are widely scattered, composed in different contexts."—Ed.]

3. Charles Darwin, *The Descent of Man, and Selection in Relation to Sex* (London: John Murray, 1871).

has appeared in modern times).[4] Darwin's letters, the correspondence with Huxley, Lyell, Hooker, Wallace, Gray, and many others, attest to a lively, informed, catholic community of scientists, much like the one Humboldt encountered in the first decade of the twentieth century.[5] *The Formation of Vegetable Mould Through the Action of Worms* (1881), which I will come to later, is a penetrating ecological study.[6] And this list omits the works on physical geography, coral reefs, and many others. For these reasons, I find the strong defense of Darwin and his methodology by Michael Ghiselin in *The Triumph of the Darwinian Method* convincing.[7]

Darwin introduced *The Origin of Species* with three quotations, the first two by the stalwart British natural theologians William Whewell and Joseph Butler, and the third by Francis Bacon, a philosopher well known to all British scientists for his criticism of the use of final causes in science. Whatever his reasons for introducing his great work with these three quotations, the effect is to say *The Origin of Species* is in a great tradition of nature study, whether that study is inspired by teleology and design or, as in the case of Bacon, by research divorced from final causes. Let us examine each one.[8]

Whewell: "But with regard to the material world, we can at least go as far as this—we can perceive that events are brought about not by insulated interpositions of Divine power, exerted in each particular case, but by the establishment of general laws."[9] Darwin was conforming here with the widespread agreement among natural theologians concerning the constancy of nature and a sureness and reliability in the operation of its laws, in opposition to miracles and the intervention of the deity in crucial periods of Earth history and the history of life—a belief that was not monopolized by theologians and clergymen but was held also by scientists such

---

4. Charles Darwin, *The Variation of Animals and Plants under Domestication*, 2 vols. (London: John Murray, 1868); George Perkins Marsh, *Man and Nature; or, Physical Geography as Modified by Human Action* (New York: C. Scribner, 1864).

5. Francis Darwin, ed., *The Life and Letters of Charles Darwin*, 2 vols. (London: J Unray, 1887; New York: D. Appleton, 1888).

6. Charles Darwin, *The Formation of Vegetable Mould Through the Action of Worms* (London: John Murray, 1881).

7. Michael T. Ghiselin, *The Triumph of the Darwinian Method* (Berkeley: University of California Press, 1969; repr., New York: Dover Publications, 2003). Page numbers in subsequent citations refer to the 2003 edition.

8. William Whewell (1784–1866) was a Cambridge scientist, philosopher, theologian, poet, and mathematician; Joseph Butler (1692–1752) was an English bishop, theologian, and philosopher; and Francis Bacon, Viscount St. Alban (1561–1626), was an English philosopher, scientist, statesman, and jurist.—Ed.

9. The quotation is from William Whewell, *Treatise III: On Astronomy and General Physics Considered with Reference to Natural Theology*, the Bridgewater Treatises: On the Power, Wisdom, and Goodness of God, as Manifested in the Creation (London, 1839).

as Georges Cuvier and Louis Agassiz.[10] The concept of the constancy of nature, its steady operations throughout time, implying a rejection of the senescence of nature, was well established in natural theology from Ray's time to the Duke of Argyll. It was also well established in science, which had shorn off final causes.[11]

Butler: "The only distinct meaning of the word 'natural' is stated, fixed or settled; since what is natural as much requires and presupposes an intelligent agent to render it so, i.e., to effect it continually or at stated times, as what is supernatural or miraculous does to effect it for once."[12] This quotation, too, allowing for an initial and special act of creation establishing nature, states the case for natural law and the constancy of nature. Darwin was well aware he was quoting a classic of English literature, the *vade mecum* of many subsequent British natural theologians. I have already discussed both of these thinkers, Butler and Whewell, at some length.[13]

Bacon: "To conclude, therefore, let no man out of a weak conceit of sobriety, or an ill-applied moderation, think or maintain, that a man can search too far or be too well studied in the book of God's word, or in the book of God's works; divinity or philosophy; but rather let men endeavour an endless progress or proficience in both."[14] The passage recognizes two great traditions: first, the Judeo-Christian likening of nature to a book meant to be read by God's creatures: among some thinkers the idea of man as a *contemplator mundi* was rejected, as Bacon by implication rejected it here, in favor of a conception of man as an active helper of God in continuing the work of the creation; and second, the empiricism of Bacon himself, which fostered the application of theoretical to applied science and the use of applied science to ameliorate the condition of mankind. The growth of knowledge of nature was intimately related to progress.[15]

---

10. Jean Léopold Nicolas Frédéric Cuvier (1769–1832), also known as Georges Cuvier, was a French naturalist, zoologist, and paleontologist. Jean Louis Rodolphe Agassiz (1807–73), also known as Louis Agassiz, was a Swiss-born biologist and geologist who in 1847 became a professor of zoology and geology at Harvard University and founder of its Museum of Comparative Zoology.—Ed.

11. See Clarence Glacken, *Traces on the Rhodian Shore: Nature and Culture in Western Thought from Ancient Times to the End of the Eighteenth Century* (Berkeley: University of California Press, 1967), the chapters on physico-theology.

12. The quotation is from Joseph Butler, *The Analogy of Religion, Natural and Revealed, to the Constitution and Course of Nature* (1736), 307.

13. Glacken notes here that Darwin was in error in citing the Butler quotation as from *Analogy of Revealed Religion*, that being the title of part 2 of that work. For Glacken's discussion of natural theologians, see *Traces on the Rhodian Shore*.—Ed.

14. The quotation is from Francis Bacon, *Of the Proficience and Advancement of Learning, Divine and Human*, book 1, 1605, Everyman Library (London: J. M. Dent and Sons, 1973), 8.

15. Glacken notes that "a continuing exposition of this philosophy appeared in another English classic, *The Religio Medici of Sir Thomas Browne*. (On nature as a book, see *Traces on the Rhodian Shore*, 203–205; Bacon, pp. 471–475; Browne, pp. 475–476.)".—Ed.

In considering the ecological implications of the Darwinian theory, one should bear in mind that the traditional basis of ecology, or the economy of nature, in natural history has been the design argument and its corollaries—special creation, adaptation to the environment as part of the creation plan, the unchanging utility of the various parts of an organism from the moment of its creation. It is true that the ecological basis, in the thought of Lamarck and Humboldt, was not in the design argument, but the controversies of Darwin's time show conclusively that it was still very much alive, whether it was applied to phylogeny or to the economy of nature. Consequently, when we read in *The Origin of Species* criticisms of the idea of special creation, we should not interpret them as pro forma rejections of an outmoded scientific outlook but as answers to live issues to refute which Darwin was obliged to adduce convincing evidence. In doing so, he showed ingenuity and resourcefulness, drawing on his vast and detailed knowledge of natural history to refute the special creationists. He also had to combat the anthropocentric argument, so deeply embedded in natural theology, although not absolutely necessary to it, because some natural theologians conceived of nature as a creation of God without reference to its usefulness to the human race.

In attacking the principle of utility and the adaptability of organic life to the environment from its creation onward, Darwin said,

> He who believes that each being has been created as we now see it, must occasionally have felt surprise when he has met with an animal having habits and structure not in agreement. What can be plainer than the webbed feet of ducks and geese are formed for swimming? Yet there are upland geese with webbed feet which rarely go near the water; and no one except Audubon has seen the frigate-bird, which has all its four toes webbed, alight on the ocean. (132)

This illustration could have been a perfect example for the natural theologian who assumed that God first made the environment and then created the webbed feet to deal with the problem of the water. Perfect mutual adaptation of organism and environment was the keystone in the arch of an ecology based on the design argument. The concepts of natural selection and the struggle for existence swept aside these neat and precise correlations, perennial favorites of the natural theologians.

> He who believes in the struggle for existence and in the principle of natural selection, will acknowledge that every organic being is constantly endeavoring to increase in numbers; and that if any one being varies ever so little, either in habits or structure, and thus gains an advantage over some other inhabitant of the same country, it will seize on the place of that inhabitant, however different that may be from its own place. Hence it will cause him no surprise that there should be geese and frigate-birds with

webbed feet, living on the dry land and rarely alighting on the water, that there should be long-toed corncrake, living in meadows instead of in swamps, that there should be woodpeckers where hardly a tree grows; that there should be diving thrushes and diving Hymenoptera, and petrels with the habits of auks. (132–33).

This is one of the clearest expositions in Darwin's writings of the shortcomings of the theory that environment, and especially climate, determines the geographic distribution of plants and animals. From an ecological point of view, Darwinian theory substituted for the environmental theory of plant and animal distribution, the theory of natural selection, and the struggle for existence, which can explain the apparent environmental anomalies that Darwin mentioned.

How about the scale of being, structural similarities in life forms, not as they bear on phylogeny but on the economy of nature?

Why, on the theory of creation, should there be so much variety and so little real novelty? Why should all the parts and organs of many independent beings, each supposedly to have been separately created for its proper place in nature, be so commonly linked together by graduated steps? Why should not Nature take a sudden leap from structure to structure? On the theory of natural selection, we can clearly understand why she should not; for natural selection acts only by taking advantage of slight successive variations; she can never take a great and sudden leap, but must advance by short and sure, though slow steps. (143–44)

In this passage, on can see the contrast between Darwin's view of the interrelationships in nature and the idea of a great chain of being in which the life forms were arranged along a scale. In Darwin's view, the interrelationships and the similarities among organisms were the result of evolution through the process of natural selection. The passage is also a clear statement of another pivotal idea of Darwin's, the precise role of natural selection. It is powerless to act unless the variations are present, a point Darwin emphasized in several letters to his correspondents and critics.[16] The economy of nature, ecological relationships among living things and their environments, he argued, are a product of evolution.

Darwin also attacked another sacred dogma, at least to many natural theologians—the aesthetic and glory-of-God theories that historically had been used to fortify the design argument. From early Christian times, it had been widely believed that life forms were created for their beauty, to please man or God or to achieve variety in nature. He countered: "Such doctrines, if true, would be absolutely fatal

---

16. Glacken in his notes provides more textual references: "To W. H. Harvey, 'More Letters of Charles Darwin,' Vol. 1, Letter 110, pp. 160–161; and *The Life and Letters of Charles Darwin*, Vol. 2, pp. 68–70, 107. In a letter to J. D. Hooker, July 13 [1861]."—Ed.

to my theory" (146). He argued further that "the sense of beauty obviously depends on the nature of the mind, irrespective of any real quality in the admired object; and that the idea of what is beautiful is not innate or unalterable. We see this, for instance, in the men of different races admiring an entirely different standard of beauty in their women" (147). This rebuttal was an adroit use of cultural relativism:

> If beautiful objects had been created solely for man's gratification, it ought to be shown that before man appeared, there was less beauty on the face of the earth than since he came on the stage. Were the beautiful volute and cone shells of the Eocene epoch, and the gracefully sculptured ammonites of the Secondary period, created that man might ages afterwards admire them in his cabinet? Few objects are more beautiful than the minute siliceous cases of the diatomaceae: were these created that they might be examined and admired under the higher powers of the microscope? (147)

Darwin's last example I have already mentioned: the inability of special creation, in Darwin's opinion, to account for the flora of oceanic islands. It could not be demonstrated that a sufficient number of the best-adapted plants and animals were intentionally created for them (305, 283).

One advantage the special creationists had over Darwin was that they had a definite beginning for the history of nature, often the exact day, or days, of the creation. In Darwinian theory, the starting point of evolution is uncertain. Evolution occurs by means of natural selection, and natural selection operates only when variations, from which it can select, are present. When in the history of nature does this process of selecting begin? If it begins with a nature full of variations, this presupposes natural selection, and nature therefore is already an ongoing concern. Did a nature prior to the production of variations exist? The natural theologians believed the Darwinian theory was consistent with initial creative acts of God because natural selection operated after the creation and not concurrently with it. How, in fact, did nature develop up to the point at which variations were possible? Darwin, so far as I know, was not scientifically concerned with the origin of life. In the last sentence of *The Origin of Species* he speaks of a view of life "with its several powers having been originally breathed by the Creator into a few forms or new ones," but what happened then? Did the original organisms multiply, according to Malthusian theory, did variations occur, with a consequent vast increase in numbers, and did natural selection then begin? Lamarck solved the difficulty by positing, without further ado, an original creation, and professed ignorance of the early periods in the history of nature.

Darwin encountered semantic difficulties in his use of "natural selection" and "nature." Did "natural selection" imply, in being contrasted with "artificial selection," that it was a conscious, directing, and intelligent agent?

> I have called this principle, by which each slight variation, if useful, is preserved, by the term Natural Selection, in order to mark its relation to man's power of selection.... We have seen [that is, in chapter 1, "Variation under Domestication"] that man by selection can produce great results, and can adapt organic beings to his own uses, through the accumulation of slight but useful variations, given to him by the hand of Nature. But Natural Selection, as we shall hereafter see, is a power incessantly ready for action, and is as immeasurably superior to man's feeble efforts, as the works of Nature are to those Art. (52)

Expressions like "hand of Nature" and "power incessantly ready for action" are metaphorical and, by themselves, do not explain much. In the extended exposition of the text, however, it was apparent what Darwin meant. Later, as if sensing the difficulty, he said,

> In the literal sense of the word, no doubt, natural selection is a false term; but who ever objected to chemists speaking of the elective affinities of the various elements? And yet an acid cannot strictly be said to elect the base with which it in preference combines. It has been said that I speak of natural selection as an active power or Deity; but who objects to an author speaking of the attraction of gravity as ruling the movements of the planets? Everyone knows what is meant and is implied by such a metaphorical expression; and they are almost necessary for brevity. So again it is difficult to avoid personifying the word Nature; but I mean by Nature, only the aggregate action and product of many natural laws, and by laws the sequence of events as ascertained by us. With a little familiarity such superficial objections will be forgotten. (64)

Whether the objections were superficial or not, they were not forgotten by George John Douglas Campbell (1823–1900), eighth Duke of Argyll, a member of the Scottish aristocracy, who held the office of Lord Privy Seal in Gladstone's administration of 1880 and whose son, the Marquis of Lorne, married Princess Louise, fourth daughter of Queen Victoria. The duke was also a pallbearer at Darwin's funeral in Westminster Abbey on April 26, 1882. "The Duke of Argyll," the only name he used as an author, is remembered, if he is remembered at all, for two books, *The Reign of Law* (1866) and *The Unity of Nature* (1884).[17] The duke was an indefatigable supporter of design and teleology in nature, and he belonged to that group of thinkers who saw in the advances of science only confirmations, not refutations, of the design argument. Both of these works were basically concerned with the relationships of science to religion: "It matters not in what department of investigation inquiry is conducted, it matters not what may be the Philosophy or

---

17. John Campbell, *The Reign of Law* (London: Strahan, 1867) and *The Unity of Nature* (New York: Putnam, 1884).

Theology of the inquirer. Every step he takes he finds himself face to face with facts which he cannot describe intelligibly either to himself or others, except by referring them to that function and power of Mind which we know as Purpose and Design."[18] For proof he turned to, of all people, Darwin! "Perhaps no illustration more striking of this principle was ever presented than the curious volume published by Mr. Darwin on the 'Fertilization of Orchids.'"[19] After summarizing Darwin's exposition of the fertilization of almost all orchids by insects carrying pollen from one flower to another, and the elaborate structure of the flower that enabled it to accomplish this, the duke, noting "Mr. Darwin's inimitable powers of observation and description," said that he used the vocabulary of design, mind, and purpose.[20]

The duke refused to accept the explanation that Darwin's language was merely "metaphorical."[21] He was fortified in his criticism by a comment Alfred Russell Wallace had made in reviewing *The Reign of Law:* "Mr. Darwin has laid himself open to much misconception, and has given to his opponents a powerful weapon against himself, by his continual use of metaphor in describing the wonderful co-adaptations of organic beings."[22]

Of all Darwin's critics from the point of view of natural theology, special creation, and design, the Duke of Argyll was in my opinion the fairest, the most incisive, and the most perceptive. His point was that Darwin often used the language of intelligent purpose in describing natural phenomena and that when necessary, he resorted to teleological explanation. In *The Unity of Nature* (1884) he returned to these claims, and particularly to the propriety of Darwin's use of metaphorical language, in light of his scientific aims.[23] The Duke of Argyll did not criticize Darwin for using the language of purpose and contrivance; he believed this was the correct language to use. What he objected to was Darwin's use of teleological language while rejecting teleology in explaining the interrelationships in nature.[24] The duke took a more ecological view of Darwin's conception of nature than many of Darwin's critics.

18. Campbell, *The Reign of Law,* 36.
19. Ibid., 36.
20. Ibid., 36, 38–39. Charles Darwin's *On the Various Contrivances by which Orchids are Fertilized by Insects* was first published by John Murray in London in 1862, the second edition in 1877.
21. Campbell, *The Reign of Law,* 36, 42–43.
22. Alfred Russel Wallace, "Creation of Law," Contributions to the Theory of Natural Selection (London: Macmillan, 1870), 264–301, at 269. [Glacken noted that "this article was reprinted, 'with a few alterations and additions,' from the *Quarterly Journal of Science,* October 1867, the year following the publication of *The Reign of Law.* The duke of Argyll quoted this in the 5th London ed., from the QJS article, 'against himself "being omitted,"' p. 41, n. 1."—Ed.]
23. Campbell, *The Unity of Nature,* 168–70.
24. Ibid., 170.

> The Correlation of Natural Forces so adjusted as to work together for the production of use in the functions, for the enjoyments, and for the beauty of Life—this is the central idea of his system; and it is an idea which cannot be worked out in detail without habitual use of the language which is moulded on our own consciousness of the mental powers by which all our own adjustments are achieved. This is what, perhaps, the greatest Observer that has ever lived cannot help observing in Nature; and so his language is thoroughly anthropopsychic.[25]

"Anthropopsychism," which the duke defined as "man-Soulism," sent me scurrying to *Webster's Third New International Dictionary of the English Language Unabridged*, where it is defined as "ascription of a soul like that of man to nature or to something that governs natural processes." The Greek-rooted word is thus kin to Ruskin's pathetic fallacy, which he deplored, but the duke saw no harm in anthropopsychism.[26]

> Seeing in the methods pursued in Nature a constant embodiment of his own intellectual conceptions, and a close analogy with the methods which his own mind recognizes as "contrivance," he rightly uses the forms of expression which convey the work of Mind. "Rightly," I say, provided the full scope and meaning of this language be not repudiated. I do not mean that naturalists should be always following up their language to theological conclusions, or that any fault should be found with them when they stop where the sphere of mere physical observation terminates. But those who seek to remodel philosophy upon the results of that observation cannot consistently borrow all the advantage of anthropopsychic language, and then denounce it when it carries them beyond the point at which they desire to stop.[27]

Finally, the Duke of Argyll mentioned what later commentators have also pointed out, the teleological character of the Darwinian theory of evolution. The concluding paragraph of *The Origin of Species* has often been quoted in this connection,[28] but the duke did not mention this. He said:

---

25. Ibid., 170–71.
26. John Ruskin coined the term "pathetic fallacy" in a chapter titled "Of the Pathetic Fallacy" in volume 3 of his *Modern Painters* (5 vols.; New York: John Wiley and Sons, 1890). Ruskin used the contemporary meaning of the word "fallacy," which was "falseness." The pathetic fallacy was committed by poets and artists when they falsely attributed emotion to physical objects, such as describing the foam of oceans as cruel, or depicted physical properties falsely, such as describing a crocus as gold when it was actually, according to Ruskin, saffron in color.—Ed.
27. Campbell, *The Unity of Nature*, 171.
28. The following is the text of the concluding paragraph: "Thus, from the war of nature, from famine and death, the most exalted object of which we are capable of conceiving, namely, the production of the higher animals, directly follows. There is grandeur in this view of life, with its several powers, having been originally breathed into a few forms or into one; and that, whilst this

The theory of Development is not only consistent with teleological explanation, but it is founded on teleology, and on nothing else. It sees in everything the results of a System which is ever acting for the best, always producing something more perfect or more beautiful than before, and incessantly eliminating whatever is faulty or less perfectly adapted to every new condition. Professor [John] Tyndall himself cannot describe this system without using the most intensely anthropopsychic language: the continued effort of animated nature is to improve its conditions and raise itself to a loftier level.[29]

The duke's remarks suggest the final lines of the *Origin*.[30]

The concepts of teleology and design in Darwin's writings, and his attitude toward them, are most puzzling to me. In *The Triumph of the Darwinian Method*, Ghiselin argued that Darwin's *The Various Contrivances by which Orchids are Fertilized by Insects* "is a metaphysical tract. It constitutes a deliberate, planned attack upon the argument from design."[31] This most interesting assertion can be placed alongside the Duke of Argyll's criticism of Darwin's use of the language of design. The duke took it seriously, but perhaps he lacked a sense of humor. Ghiselin cited a letter from Darwin, written at Down and dated September 21, 1861, to John Murray, his publisher. Of his book, Darwin wrote, "Like a Bridgewater treatise, the chief object is to show the perfection of the many contrivances in Orchids."[32] According to Ghiselin, "The book is a sort of biological *Candide*, which, albeit with the greatest restraint, holds up the very idea of organic design to ridicule and contempt. There are reasons to think, indeed, that it was written as a deliberate satire on the *Bridgewater Treatises*."[33] Ghiselin may be right, but Darwin's correspondence on the orchid book and his comments on reviews of it certainly give me the impression that it was not meant as a "metaphysical tract," a satire, or a parody. To complicate matters further, Francis Darwin, his son, remarked, "One of the greatest services rendered by my father to the study of Natural History is the revival of Teleology. The evolutionist studies the purpose or meaning of organs with the zeal of the older Teleology, but with far wider and more coherent purpose."[34] Francis Darwin

---

planet has gone cycling on according to the fixed law of gravity, from so simple a beginning endless forms most beautiful and most wonderful have been, and are being, evolved."—Ed.

29. Campbell, *The Unity of Nature*, 171.

30. Glacken notes, "Tyndall was a physicist who also wrote many popular works on science. The source of the quotation is given only as *Belfast address*."—Ed.

31. Ghiselin, *Darwinian Method*, 135.

32. F. Darwin, ed., *Life and Letters*, 2:441.

33. Ghiselin, *Darwinian Method*, 136.

34. F. Darwin, ed., *Life and Letters*, 2:430.

himself was a botanist and plant physiologist. But who should have known what "teleology" in nature implied at that time? There is no hint that Darwin thought the work on orchids was a parody on design. If it were, Asa Gray, sensitive to criticisms of design, would surely have caught it.[35]

I have dwelt on metaphorical language, the personification and teleology of nature in Darwin's thought, and on the criticism of the Duke of Argyll because these subjects are excellent illustrations, perceived in the nineteenth century both by Darwin and by the duke, of the close relationship of the concepts of nature to language. This is the second time we have seen it. Humboldt, in an entirely different context, also saw the critical role of language in constructing concepts of nature. The problems the Duke of Argyll posed as difficulties in the Darwinian theory of nature to my knowledge have never been solved; they seem to have been ignored. The subject is still an open one. Most students recognize metaphysical, metaphorical, or symbolic overtones in the vocabulary of their discipline, and ecological theory following Darwin has many words of a metaphorical and teleological character, such as succession, climax, community, holism, cooperation, unconscious cooperation, equilibrium, and balance.

Before discussing the ecological aspects of Darwin's theory of evolution, I wish to mention another idea that has been influential in interpreting Darwin's philosophy of nature: that natural selection, the struggle for existence, is based on an analogy taken from the social world, from competition in modern society as manifested in the English factory system, perhaps even inspired by Hobbes's "war of all against all": "It is remarkable, how Darwin recognizes among beasts and plants his English society with its division of labour, competition, opening up of new markets, 'inventions,' and the Malthusian 'struggle for existence.' It is Hobbes's 'bellum omnium contra omnes,' and one is reminded of Hegel's *Phenomenology*, where civil society is described as a 'spiritual animal kingdom,' while in Darwin the animal kingdom figures as civil society."[36] To me, this analogy is misguided and ignores the history of natural history in Western civilization. Darwin's *The Origin of Species* can be regarded as the culmination of research in natural history lasting more than 150 years, from John Ray and Linnaeus to Buffon, Banks, Lamarck, Cuvier, Geoffroy Saint-Hilaire, and many others. The observation of a struggle for existence in

---

35. Glacken here references the following Darwin correspondence gleaned from *Life and Letters*: his letter of July 12, 1860, to Hooker, 2:437–38; his reaction to reviews, pp. 444–45; his letter of June 10, 1862, to Asa Gray, pp. 445–47. Asa Gray, a professor of botany at Harvard, was one of the leading American botanists of the nineteenth century.—Ed.

36. Glacken attributes this quotation to Alfred Schmidt, *The Concept of Nature in Marx*, 46. Source: Karl Marx and Friedrich Engels, *Selected Correspondence* (London, n.d.), 156–57. Letter of June 18, 1963.—Ed.

nature, without resorting to analogies from social institutions, is very ancient; it can be found in Lucretius.[37] In modern times, natural theologians were much impressed by predation in nature, especially by the theological problem posed by the larger carnivores; it was clearly necessary to include them within the design and then to make their inclusion convincing.

The acceptance of the analogy would not be surprising if one were only at home in the political, social, and philosophical thought of the eighteenth and nineteenth centuries, and if one were innocent of the writings of the natural historians. Studies have been made of the history of the idea of natural selection, competition and warfare in nature, ideas that have had little to do with human warfare, competition, or industrialism. Some writers, including Saint-Pierre and Buffon, in fact used the life of animals in nature as a means of criticizing contemporary human society. Darwin may not have been a profound student of the history of natural history, but his books, notebooks, and correspondence are full of evidence that his ideas were not based on crude analogies with human society. The correspondence with Hooker, Lyell, Huxley, and many others shows—indeed, if this has to be shown—a deep immersion in natural history, not only in its bearings on evolution but in and for itself, as his studies of earthworms and the movements of plants demonstrate.

Although I reject an explanation of Darwin's philosophy of nature as being derived from nineteenth-century political and social philosophy, I do not argue that no influence came from the social world. It has often been pointed out that there were three building blocks of the Darwinian concept of nature: uniformitarianism, artificial selection, and the principle of population. The first came from geology, in part from Hutton, then Playfair and Lyell, and the last two from the social world. The activities of plant and animal breeders suggested, through the availability of varieties, to Darwin the mechanism of natural selection, but plant and animal breeding was not a political and social theory but a highly developed applied science, as Darwin's comments on plant and animals breeders and his quotations from their writings show. It is true that the argument was based on an analogy of sorts that is, if weak human beings can bring about such wondrous changes in plants and animals by breeding, a much more powerful nature can effect infinitely vaster changes.

The influence of Malthus's doctrine of population is crystal-clear. I cannot understand those who tone down or deemphasize the influence of Malthus on Darwin, in view of his explicit statements in his autobiography and in *The Origin of Species*. Although it is true that Malthus's principle of population was concerned with human societies; he believed the principle applied to all life. The following

---

37. Glacken, *Traces on the Rhodian Shore*, 72–73.

passage from Malthus's *First Essay* could well have been in chapter 3 of *The Origin of Species*:

> Through the animal and vegetable kingdoms, nature has scattered the seeds of life abroad with the most profuse and liberal hand. She has been comparatively sparing in the room and the nourishment necessary to rear them. The germs of existence contained in this spot of earth, with ample food, and ample room to expand in, would fill millions of worlds in the course of a few thousand years. Necessity, that imperious all pervading law of nature, restrains them within the prescribed bounds. The race of plants, and the race of animals shrink under this great restrictive law. And the race of man cannot, by any efforts of reason, escape from it.[38]

A "cause" that had hitherto been neglected in the study of human history, Malthus said, "is the constant tendency in all animated life to increase beyond the nourishment prepared for it."[39] It is possible to argue that Malthus derived the principle of population not from human society but from the natural world, limiting himself in the essay to the former. Malthus, however, did not instruct Darwin in the struggle for existence in nature, as is abundantly clear from the following often-quoted passage from Darwin's *Autobiography*. Malthus was more of a catalyst who influenced Darwin to look at the struggle from a new and different viewpoint.

> In October 1838, that is, fifteen months after I had begun my systematic enquiry, I happened to read for amusement "Malthus on Population," *and being well prepared to appreciate the struggle for existence which everywhere goes on from long-continued observation of the habits animals and plants,* it at once struck me that under these circumstances favourable variations would tend to be preserved, and unfavourable ones to be destroyed. The result of this would tend to be the formation of new species. Here then I had at last got a theory by which to work; but I was so anxious to avoid prejudice, that I determined not for some time to write even the briefest sketch of it. . . .[40]

Although I do not deny for a moment the interpenetration of ideas in the social and the natural world, the added perspective from the history of natural history is needed in judging this question in the thought of Darwin. Some of the roots of Darwin's ideas about nature certainly lie in the past, in part connected to the long British tradition of pursuing natural history, represented by such classics as Gilbert

---

38. Malthus, *Population: The First Essay* (Ann Arbor: University of Michigan Press, 1959), vol. 1, chap. 1, p. 5.—Ed.

39. Malthus, *An Essay on Population*, 1:5.

40. F. Darwin, ed., *Life and Letters*, 1:68. [Glacken notes that the emphasis was added by him.—Ed.]

White's *The Natural History of Selborne*.[41] Little in Darwin's life or in his intellectual history suggests much influence from political and social theory; in *The Descent of Man*, he showed a wide reading of prehistorians and anthropologists, but they were not concerned with political questions.

The Ecological Implications of Darwin's Theory

The preceding discussions have prepared the way, I hope, for an exposition of Darwin's ecological thought. The theory of evolution by means of natural selection is indissolubly linked with the concept of the interrelationships of organisms with one another and with their physical environments because of the immense variety and number of organisms living together in equally diverse associations. If one reads *The Origin of Species* from this viewpoint, it is striking to see how many of Darwin's illustrations come from everyday observation of ecological relationships in nature. Characteristically, Darwin is modest about his own knowledge and acutely aware of what is unknown.

> No one ought to feel surprised at much remaining as yet unexplained in regard to the origin of species and varieties, if he make due allowance for our profound ignorance in regard to the mutual relations of the many beings which live around us. Who can explain why one species ranges widely and is very numerous, and why another allied species has a narrow range and is rare? Yet these relations are of the highest importance, for they determine the present welfare and, as I believe, the future success and modification of every inhabitant of this world. Still less do we know of the mutual relations of the innumerable inhabitants of the world during the many past geological epochs in its history. (13)[42]

Darwin criticized the doctrine of special creation from an ecological point of view and assumed that observed interrelationships in nature are products of evolution: "Almost every part of every organic being is so beautifully related to its complex conditions of life that it seems as improbable that any part should have been suddenly produced perfect, as that a complex machine should have been invented by man in a perfect state" (38). Chapter 3 of *The Origin of Species* is titled "Struggle for Existence," and it is a pity that this expression—grim, striking, easily quoted—should have been used as a synonym for ecological relationships in one of the most important chapters of the book. It would have been much better if Darwin had

---

41. Gilbert White (1720–93) was an English priest, naturalist, and ornithologist, and author of *The Natural History and Antiquities of Selborne, in the County of Southampton* (London: White, Cochrane, and Co., 1813). The book has been continuously in print since then.—Ed.

42. The thought is repeated on p. 96 of the *Origin*.

used the old-fashioned expression, "the economy of nature." One might reply that I put too much emphasis on a word, but the simple truth is that "Principles of Ecology" would have been a much better designation for a chapter on the struggle for existence. The trouble with the term is that it was too narrow, too emotional, and encouraged analogy, whose quick rise, by returning Malthusianism in harsher language to society, supplied ready-made justifications for imperialism, warfare, the oppression of primitive peoples, and racism. When a new expression is coined, volunteers are always ready to import it into another discipline, and many more follow, thinking themselves in the avant-garde when they are only on a bandwagon. To Darwin, however, the struggle for existence was synonymous with the concept of "a web of complex relations" (59).

I wish, therefore, to point out some passages in Darwin's writings that have this ecological character and that inspired, directly or indirectly, some ecological studies in the latter part of the nineteenth and the early part of the twentieth centuries. "Adaptation" is certainly one of the most important of these concepts. It has a long history, and it was a key idea in the design argument. The notion that a plant or an animal is adapted to its physical environment, that its physical constitution, by design, makes this adaptability possible, is among the oldest observations of natural history. It is prominent in natural histories based on the design argument because a reasonable, just, and kind God would only design and bring a creation into being whose constituent parts were adapted to one another. With Darwin, however, adaptation was not the result of design or of special creation but of evolution. Adaptation has probably had such an important role in the history of ideas concerning the natural world because it can be easily observed by a trained or an untrained eye.

> How have all those exquisite adaptations of one part of the organization to another part, and to the conditions of life, and of one organic being to another being, been perfected? We see these beautiful co-adaptations most plainly in the woodpecker and the mistletoe; and only a little less plainly in the humblest parasite which clings to the hairs of a quadruped or feathers of a bird; in the structure of the beetle which dives through the water; in the plumed seed which is wafted by the gentlest breeze; in short, we see beautiful adaptations everywhere and in every part of the organic world. (51)

Such observations Darwin put to work in describing the mechanisms of natural selection.

Although "the struggle for existence" is prominent in his thought, it seems subordinated to a broader ecological concept, along with a reminder that appearances can be deceiving.

> Nothing is easier than to admit in words the truth of the universal struggle for life, or more difficult—at least I have found it so—than constantly to bear this conclusion in mind. Yet unless it be thoroughly engrained in the mind, the whole economy of nature, with every fact on distribution, rarity, abundance, extinction, and variation, will be dimly seen or quite misunderstood. We behold the face of nature bright with gladness, we often see superabundance of food; we do not see or we forget, that the birds which are idly singing round us mostly live on insects or seeds, and are thus constantly destroying life; or we forget how largely these songsters, or their eggs, or their nestlings, are destroyed by birds and beasts of prey; we do not always bear in mind, that, though food may be now superabundant, it is not so at all seasons of each recurring year. (52)

In a passage reminiscent of a favorite theme of Lamarck, that little attention had been given in zoology to the invertebrates, Darwin spoke of the universality of the reproductive power of nature and its application, despite appearances to the contrary, to all forms of life.

> In a state of nature almost every full-grown plant annually produces seed, and amongst animals there are very few which do not annually pair. Hence, we may confidently assert, that all plants and animals are tending to increase at a geometrical ratio,—that all would rapidly stock every station [a word in common use at this time for "habitat"] in which they could anyhow exist,—and that this geometrical tendency to increase must be checked by destruction at some period of life. Our familiarity with the larger domestic animals tends, I think, to mislead us: we see no great destruction falling on them, but we do not keep in mind that thousands are annually slaughtered for food, and that in a state of nature an equal number would have somehow to be disposed of. (54)

The idea of an "economy of nature" assumes in Darwin's thought a dynamic aspect; it is not the natural environment of the geographic environmentalists of the eighteenth and nineteenth centuries, which passively but effectively molds culture and natural character. The natural environment is dynamic; though in ways unlike Lamarck's conception that the environment acts directly on life forms. The elements of this dynamism are the checks on the vast reproductive power of life. Darwin used the term in the same sense that Malthus used it.

> The causes which check the natural tendency of each species to increase are most obscure. Look at the most vigorous species; by as much as it swarms in numbers, by so much will it tend to increase still further. We know not exactly what the checks are even in a single instance. Nor will this surprise any one who reflects how ignorant we are on this head, even in regard to mankind, although so incomparably better known than any other animal. (55)

The section "Nature of the Checks to Increase," in chapter 3 of the *Origin*, is evidence of the effectiveness of Darwin's observation in documenting his theory because his discussion of the checks on the increase of life obviously owes much to the interests and activities of the traditional British natural historian. I do not believe he needed the inspiration of Malthus, although the latter's discussion of checks on the reproductive power of human beings may well have been stimulating. The checks in the natural world could come from the struggle for existence between individuals of the same species, from interspecies competition, or from the physical environment, especially but indirectly from climate. Speaking of these checks, Darwin said:

> Eggs or very young animals seem generally to suffer most, but this is not invariably the case. With plants there is a vast destruction of seeds, but from some observations that I have made it appears that seedlings suffer most from germinating in ground already thickly stocked with other plants. Seedlings, also, are destroyed in vast numbers by various enemies; for instance, on a piece of ground three feet long and two wide, dug and cleared, and where there could be no choking from other plants, I marked all the seedlings of our native weeds as they came up, and out of 357 no less than 295 were destroyed, chiefly by slugs and insects. (55)

Occasionally in his ecological discussions, Darwin introduced the human influence on the web of complex relations: "If turf which has long been mown, and the case would be the same with turf closely browsed by quadrupeds, be let to grow, the more vigorous plants gradually kill the less vigorous" (55). Predation is another significant check on population increase in nature, and the average number of a species might be determined not by food but by being prey for other animals.

> Thus, there seems to be little doubt that the stock of partridges, grouse, and hares on any large estate depends chiefly on the destruction of vermin. If not one head of game were shot during the next twenty years in England, and at the same time, if no vermin were destroyed, there would, in all probability, be less game than at present, although hundreds of thousands of game animals are now annually shot. (55–56)

Epidemics and parasitism are further checks on population growth, and the fewness of individuals of a species compared with their predators' numbers can also become a check on their increase (57).

It is amazing how vital climate has been in the history of social and biological thought in Western civilization. In the social world, ideas of the influence of climate on culture began with Hippocrates, continued with several others in antiquity, with Albert the Great in the Middle Ages, Bodin in the Renaissance, and Montesquieu in the eighteenth century.[43] Climate has also had a vitally important role in the history of biological thought in the literature relating to climatic influences on the effect of seasonal change on biotic communities, the distribution of plants and animals; in climatic changes in the Earth's history and their effect on the structure and evolution of plants and animals; and in climate as an exterminator of life. Darwin considered climate as a check on population increase, but, consistent with his theory of natural selection, he was inclined, without denying a direct influence, to emphasize its indirect effects.

> Climate plays an important part in determining the average number of a species, and periodical seasons of extreme cold or drought seem to be the most effective of all checks. I estimated, (chiefly from the greatly reduced numbers of nests in the spring) that the winter of 1854–5 destroyed four-fifths of the birds in my own grounds; and this is a tremendous destruction when we remember that ten per cent is an extraordinarily severe mortality from epidemics with man. The action of climate seems at first sight to be quite independent of the struggle for existence; but in so far as climate chiefly acts in reducing food, it brings on the most severe struggle between the individuals, whether of the same or of distinct species, which subsist on the same kind of food. (56)

Darwin also carefully distinguished among the various influences of climate on the struggle for existence. Those effects differ with latitude and altitude. Again he rejected direct climatic explanations of the geographic distribution of plants and animals and direct climatic influences on the chances for survival.

> When we travel from south to north, or from a damp region to a dry, we invariably see some species gradually getting rarer and rarer, and finally disappearing; and the change of climate being conspicuous, we are tempted to attribute the whole effect to its direct action. But this is a false view; we forget that each species, even where it most abounds, is constantly suffering enormous destruction at some period of its life, from enemies or from competitors for the same place or food; and if these enemies or competitors be in the least degree favored by any slight change of climate, they will increase in numbers; and as each area is already fully stocked with inhabitants, the other species must decrease. When we travel southward and see a species decreasing in numbers, we may feel sure that the cause lies quite as much in other species being favored, as in this one being hurt. So it is when we travel northward, but in somewhat lesser degree, for the number of species of all kinds, and therefore of competitors, decreases northwards; hence in going northwards, or in ascending a mountain, we far oftener meet with stunted forms,

---

43. For a history of ideas about the environment in Western civilization as they have related to culture, see Glacken, *Traces on the Rhodian Shore*, chaps. 2, 6, 9, 12, 13.

due to the *directly* injurious action of climate, than we do in proceeding southwards or in descending a mountain. When we reach the Arctic regions, or snow-capped summits, or absolute deserts, the struggle for life is almost exclusively with the elements. (56)

The heart of chapter 3, and probably the most concise and cogent ecological exposition in the writings of Darwin, is the section titled "Complex Relations of all Animals and Plants to each other in the Struggle for Existence" (57–58). This is one of the most familiar and most often quoted passages in the *Origin*, especially the cats-to-clover chain, and is a classic exposition of an ecological view of nature based on evolution and not on design.[44] This passage also illustrates a conspicuous characteristic of Darwin's thought and research—the frequency with which he uses observations taken from contemporary life and from personal experience in support of broad and fundamental generalizations. His writings frequently mention conversations with friends, neighbors, his sons, or a scientific colleague. Indeed, many of Darwin's examples involve, directly or indirectly, the participation of human beings in the complex web.[45] The "web of complex relations" passage also provides a way of looking at human intrusions in the natural world. The human activities described by Darwin did not take place as events on a stage; they were intimately involved in the web. The idea of a web (a biocenose, a biotic commu-

---

44. The "cats-to-clover chain" is encapsulated in the following quotation from the same page: "I am tempted to give one more instance showing how plants and animals, most remote in the scale of nature, are bound together by a web of complex relations. I shall hereafter have occasion to show that the exotic Lobelia fulgens, in this part of England, is never visited by insects, and consequently, from its peculiar structure, never can set a seed. Many of our orchidaceous plants absolutely require the visits of moths to remove their pollen-masses and thus to fertilise them. I have, also, reason to believe that humble-bees are indispensable to the fertilisation of the heartsease (Viola tricolor), for other bees do not visit this flower. From experiments which I have tried, I have found that the visits of bees, if not indispensable, are at least highly beneficial to the fertilisation of our clovers; but humble-bees alone visit the common red clover (Trifolium pratense), as other bees cannot reach the nectar. Hence I have very little doubt, that if the whole genus of humble-bees became extinct or very rare in England, the heartsease and red clover would become very rare, or wholly disappear. The number of humble-bees in any district depends in a great degree on the number of field-mice, which destroy their combs and nests; and Mr H Newman, who has long attended to the habits of humble-bees, believes that 'more than two thirds of them are thus destroyed all over England.' Now the number of mice is largely dependent, as every one knows, on the number of cats; and Mr Newman says, 'Near villages and small towns I have found the nests of humble-bees more numerous than elsewhere, which I attribute to the number of cats that destroy the mice.' Hence it is quite credible that the presence of a feline animal in large numbers in a district might determine, through the intervention first of mice and then of bees, the frequency of certain flowers in that district!"—Ed.

45. For a more complete treatment of this subject by Glacken, see "This Growing Second World within the World of Nature, in *Man's Place in the Island Ecosystem*, ed. F. R. Fosberg (Honolulu: Bishop Museum Press, 1963), 75–95; references on 89–90.—Ed.

nity, an ecosystem) becomes an invaluable tool in understanding the relationship of culture to environment.

Most students of Darwin's scientific method and philosophy of nature have stressed the theory of evolution and the criticisms of special creation, design, and final causes in nature. Others, however, such as Ernst Haeckel, saw the pertinence of his thought to ecology. J. Arthur Thomson, a Scottish biologist, the Regius Professor of Natural History at the University of Aberdeen and the author of several popular books on science and religion, was aware of the historical continuity in this interpretation of the natural world.

> Naturalists, in the true sense, who study the life of living creatures in nature have always been distinguished by a keen perception of the inter-relations of things. Whether we take Gilbert White [*The Natural History of Selborne*] as representing the old school, or W. H. Hudson as representing the new, we get from their observations the same impression of nature as a liberating system, most surely and subtly interconnected but it seems just to say that no naturalist, before or since, has come near Darwin in his realization of the web of life, in his clear vision and picture of the vast system of linkages that penetrates throughout the animate world.[46]

Although he did not elaborate on the point, Thomson recognized the relationship between ecology and evolution. "There is a changing pattern in the web, becoming more complex as the ages pass; and *This is evolution.*"[47]

I wish to add few more illustrations from the works of Darwin because, although they further illustrate the same theme of interrelationships, they also reveal the scope and breadth of Darwin's ecology. In a discussion of "organs of little apparent importance, as affected by natural selection," he wrote,

> The tail of the giraffe looks like an artificially constructed fly-flapper; and it seems at first incredible that this could have been adapted for its present purpose by successive slight modifications, each better and better fitted, for so trifling an object as to drive away flies.... It is not that the larger quadrupeds are actually destroyed (except in some rare cases) by flies, but they are incessantly harassed and their strength reduced, so that they are more subject to disease, or not so well enabled in a coming dearth to search for food, or to escape from beasts of prey. (144)

Like Lamarck, Darwin realized the importance of including the invertebrates in the economy of nature, and his writings, as in his work on orchids, frequently discuss the role of insects in the web of life.

---

46. J. Arthur Thomson, *Darwinism and Human Life* (New York: Henry Holt, 1910), 45.
47. Ibid., 46–47.

Darwin was careful not to be dogmatic, or to ascribe to natural selection alone the mechanism by which the complex web comes into being. Sexual selection may also be at work, and organs developed for one need might gradually change to suit another.

> If green woodpeckers alone had existed, and we did not know that there were many black and pied kinds, I dare say that we should have thought that the green colour was a beautiful adaptation to conceal this tree-frequenting bird from its enemies; and consequently that it was a character of importance, and had been acquired through natural selection; as it is, the colour is probably in chief part due to sexual selection. A trailing palm in the Malay Archipelago climbs the loftiest trees by the aid of exquisitely constructed hooks clustered around the ends of the branches, and this contrivance, no doubt, is of the highest service to the plant; but as we see nearly similar hooks on many trees which are not climbers, and which, as there is reason to believe from the distribution of the thorn-bearing species in Africa and South America, serve as a defense against browsing quadrupeds, so the spikes on the palm may at first have been developed for this object, and subsequently have been improved and taken advantage of by the plant, as it underwent further modification and became a climber. (145)

The beauty of plant and animal life clearly attracted Darwin; this appreciation appears in casual asides in his writings, and it could become, as in his work on orchids, an inspiration. In considering beauty in flowers as it related to natural and sexual selection, he found it noteworthy that insects have a vital role in ecological relationships.

> Flowers rank amongst the most beautiful productions of nature; but they have been rendered conspicuous in contrast with green leaves, and in consequence at the same time beautiful, so that they may be easily observed by insects. . . . Several plants habitually produce two kinds of flowers; one kind open and coloured so as to attract insects; the other closed, not coloured, destitute of nectar, and never visited by insects. Hence we may conclude that, if insects had not been developed on the face of the earth, our plants would not have been decked with beautiful flowers, but would have produced only such poor flowers as we see on our fir, oak, nut and ash trees, on grasses, spinach, docks, and nettles, which are all fertilized through the agency of the wind. (147–48)

Darwin added, however,

> that a great number of male animals, as all our most gorgeous birds, some fishes, reptiles, and mammals, and a host of magnificently coloured butterflies, have been rendered beautiful for beauty's sake; but this has been effected through sexual selection, that is, by the more beautiful males having been continually preferred by the females,

> and not for the delight of man. So it is with the music of birds. We may infer from all this that a nearly similar taste for beautiful colours and for musical sounds runs through a large part of the animal kingdom. (148)

His concept of beauty is puzzling; he seems to believe that one sense of beauty permeates all life, that human standards of beauty are also those of beasts, birds, and insects. Or am I missing something?

One of the enduring questions about the design argument as it applied to the natural world, and granting that plants and animals living together necessarily stand in some relation to one another, has been whether one life form might be created for the benefit of another. The question is inherent in the reality of predation and in the concept of the food chain. In the ancient world, Aristotle had considered this question and answered it, with one minor exception, in the negative.[48] Darwin adapted a position similar to Aristotle's. Although he did not mention any names, apparently the view that one life form may exist for the sake of another was held by many natural historians of his day.

> Natural selection cannot possibly produce any modification in a species exclusively for the good of another species; though throughout nature one species incessantly takes advantage of, and profits by, the structures of others.... If it could be proved that any part of the structure of any one species had been formed for the exclusive good of another species, it would annihilate my theory, for such could not have been produced through natural selection. Although many statements may be found in works on natural history to this effect, I cannot find even one which seems to me of any weight. (148)

If I interpret this passage correctly, Darwin lays the groundwork, based on evolutionary theory, for studying predation and the food chain.

Many readers of *The Origin of Species* have been particularly impressed by chapters 12 and 13, which have to do with geographic distribution. They are the culmination of research and interests that were already apparent in eighteenth- and nineteenth-century natural history. Buffon was much interested in the geographic distribution of animals, believing it was controlled largely by environment. Humboldt is universally acknowledged to be the founder of plant geography. The Danish botanist Joakim Frederik Schouw built on Humboldt's work; his book, published in

---

48. W. D. Ross, *Aristotle* (London: Methuen, 1923), 125. See also *Traces on the Rhodian Shore*, 48–49. [Glacken notes: "Remarking that Aristotle's teleology is an 'immanent' teleology, Ross added that 'The end of each species [in Aristotle's thought] is internal to the species; its end is simply to be that kind of thing, or, more definitely, to grow and reproduce its kind, to have sensation, and to move, as freely and efficiently as the conditions of its existence—its habitat, for instance—allow.'"—Ed.]

Danish in 1822, had a wider circulation when he translated it into German.[49] Finally, in 1855, Alphonse de Candolle, son of the Swiss botanist August de Candolle, published the *Géographie botanique raisonée*.[50] Darwin consulted it and corresponded with him. And, at the other end, Alfred Russell Wallace, in the preface to his two-volume work, *The Geographical Distribution of Animals*, expressed the hope "that I have made some approach to the standard of excellence I have aimed at—which was, that my book should bear a similar relation to the eleventh and twelfth chapters of the 'Origin of Species,' as Mr. Darwin's 'Animals and Plants under Domestication' does to the first chapter of that work."[51]

One of Darwin's basic assumptions was that environmental conditions alone could not account for the origin of species; natural selection was more fundamental. His objection to environmental explanations was the basis for rejecting Lamarck's ideas and was maintained despite Darwin's acceptance of the Malthusian principle (a theory of environmental limitation) and his own discussion in chapter 3 of *The Origin of Species* of checks on population growth in nature, many of which were environmental, especially climatic. Later in life he retreated from this position. In a letter to Moritz Wagner dated October 13, 1876, he wrote:

> In my opinion the greatest error which I have committed, has been not allowing sufficient weight to the direct action of the environment, *i. e.* food, climate, &c., independently of natural selection. Modifications thus caused, which are neither of advantage nor disadvantage to the modified organism, would be especially favoured, as I can now see chiefly through your observations, by isolation in a small area, where only a few individuals lived under nearly uniform conditions.[52]

Moritz Wagner published *Die Darwinische Theorie und das Migrationsgesetz der Organismen* in 1868; although he had been interested in species and their distribution before reading *The Origin of Species,* Darwin's book inspired him to supplement the *Origin* and to add materials that, he thought, Darwin had not sufficiently considered.[53] Wagner emphasized migration and isolation as important factors in the or-

---

49. Frederik Schouw, *Grundzüge einer allgemeinen Pflanzengeographie* (Berlin, 1823).

50. Alphonse de Candolle, *Géographie botanique raisonnée; ou, Exposition des faits principaux et des lois concernant la distribution géographique des plantes de l'époque actuelle*, 2 vols. (Paris: Masson, 1855).

51. Alfred Russell Wallace, *The Geographical Distribution of Animals, With a Study of the Relations of Living and Extinct Faunas and Elucidating the Past Changes of the Earth's Surface*, 2 vols. (New York: Harper and Brothers, 1876), 1:xv.

52. F. Darwin, ed., *Life and Letters*, 2:338. [Glacken notes: "for continuing uncertainties, see his letter of July 19, 1881 to K. Semper, ibid., Vol. 2, pp. 516–517."—Ed.]

53. Moritz Wagner (1813–1887) was a German explorer, collector, geographer, and natural historian who developed the idea that geographic isolation could play a key role in speciation.

igin of species. Darwin in his studies of geographic distribution had certainly also recognized them. In reply to Wagner's criticism, Darwin wrote, in the same letter, "It would have been a strange fact if I had overlooked the importance of isolation, seeing that it was such cases as that of the Galapagos Archipelago, which chiefly led me to study the origin of species."[54]

"Man's Place in Nature" Reconsidered...

"Man's place in nature" is an expression that permeates the nature literature of Western civilization. It invariably has appeared, whether in religion, science, or philosophy, in conceptions of the natural world, cosmic or terrestrial. The reasons for its constant recurrence are the obvious kinship of human with other forms of life in the cycle of birth, growth, maturity, and death; the equally obvious difference between human beings and the higher animals; and the corollary question of whether the difference is a qualitative or a quantitative one. The main strength of the qualitative interpretation has been the Judeo-Christian belief that the human race is a unique creation and in God's own image. Secular conceptions have also acknowledged a great gulf between human and animal life, stressing the possession of the hand, the eye, erect carriage, and a mind capable of reason and memory, thus of the accumulation of knowledge in books, art, and music, capable of love, and capable also of retrieval. The question of man's place in nature is a persistent theme in Darwin's thought. In this section I explore this theme by broadening it to include Darwin and two of his most prominent nineteenth-century evolutionist contemporaries, Huxley and Wallace, both because they had something to say on the subject and because they were intensely interested in it themselves.

The Origin of Species was not concerned with the human race, or with its evolution. On Linnaeus's statement that no one had yet been able to discover any characteristic positive enough to authorize separating man from the ape, Darwin had this to say:

> What! the being who has surveyed ('measured') the earth and the heavens, who has analyzed the spectrum (*décompose la lumière*), who has invented languages, who has constructed all these beautiful buildings, given life to all these statues; the being who

---

His work was also an important influence on Friedrich Ratzel. The work cited by Glacken has an English translation: M. Wagner. *The Darwinian Theory and the Law of the Migration of Organisms*, trans. J. L. Laird, (London: Edward Stanton, 1873). Glacken also refers to Hanno Beck, "Moritz Wagner in der Geschichte der Geographie" (diss., University of Marburg, 1951). On Wagner's influence on Ratzel, he refers to Johannes Steinmetzler, *Die Anthropogeographie Friedrich Ratzels und ihre ideengeschichtlichen Wurzeln* (Bonn: Bonner Geographische Abhandlungen, 1956), 19:93–99.—Ed.

54. F. Darwin, ed., *Life and Letters*, 2:338.

has subdued steam and made it a faithful executor of his desires, the being who thinks and foresees, the being endowed with reason, differs only from the ape by having a higher degree of intelligence! The day on which you wrote these lines, honest Linnaeus, you were no doubt complaining to yourself about one of your fellows (*semblables*) and it is in this way that you have been avenged.[55]

Darwin believed that the same processes had operated in human evolution as in that of other forms of life.[56] Variations were also induced by the same causes and laws as governed other forms of life. Early men, in their widespread migrations had been exposed to most diversified conditions "before they reached their present home." The laws of population growth (that is, the Malthusian principle of population) must have applied to man as to other forms of life, and man was involved in struggle for existence and subject to the operation of natural selection. Taking his cue from observations of present-day primitive peoples, Darwin postulated, without giving any precise time for it, the early dominance of man over the rest of life.

> Man in the rudest state in which he now exists is the most dominant animal that has ever appeared on this earth. He has spread more widely than any other highly organized form: and all others have yielded before him. He manifestly owes this immense superiority to his intellectual faculties, to his social habits, which lead him to aid and defend his fellows, and to his corporeal structure. The supreme importance of these characters has been proved by the final arbitrament of the battle for life. (431)

Although he did not pursue the thought, Darwin left the way open for inquiry into the role of early man in modifying the natural world. Lyell considered the possible role of human agency in the extinction of prehistoric animals. The subject today, of course, has become a major field of research. Presumably still relying on ethnological accounts of primitive peoples, Darwin described early men as active, skilled technologists with the ability directly or indirectly to interfere in the natural world through the use of weapons, traps, and fire. He described the discovery of fire as "probably the greatest ever made by man, excepting language," dating from before the dawn of history (432). Darwin's analysis of the skills and activities of early man should be read in the light of prevailing attitudes of his time because it might seem today obvious and conventional. In the nineteenth century, the widely accepted view of primitive peoples was that they were little more than plastic beings molded by their environment and, like the rest of life, inseparable from it. The realization of the skills of primitive peoples, their ability to modify nature, and their detailed and accurate knowledge of environment (such as knowledge

---

55. I have been unable to trace the Darwin work from which Glacken quotes.—Ed.
56. Darwin, *The Descent of Man,* chap. 1. Subsequent page citations are given in the text.

of useful plants, watering holes in the desert, and the habits of animals) is comparatively recent. "These several inventions, by which man in the rudest state has become so pre-eminent, are the direct results of the development of his powers of observation, memory, curiosity, imagination, and reason. I cannot, therefore, understand how it is that Mr. Wallace maintains that 'natural selection could only have endowed the savage with a brain a little superior to that of an ape'" (432). Darwin was referring to an article by Wallace in the *Quarterly Review*, April 1869;[57] Wallace's drawing on his experience with primitive peoples to support his belief that the higher human faculties could not have been acquired by natural selection will be discussed shortly.

Like countless previous thinkers, including natural theologians from ancient times to the nineteenth century, Darwin was much impressed by the uniqueness of the human hand and the consequent power it conferred on man. "Man could not have attained his present dominant position in the world without the use of his hands, which are so admirably adapted to act in obedience to his will. Sir C. Bell in *The Hand* [in the Bridgewater Treatises] insists that 'the hand supplies all instruments, and by its correspondences with the intellect gives him universal dominion'" (434).[58] Darwin believed there was no fundamental difference between man and the higher mammals in mental faculties.[59] Unlike many authors, including Huxley, he did not single out language as the most important unique characteristic of the human species. Animals possessed in a "rude and incipient degree" the power of forming general concepts (464). The superiority of man lay in another direction: "I fully subscribe to the judgment of those writers who maintain that of all the differences between man and the lower animals, the moral sense or conscience is by far the most important" (471).[60] Darwin quoted a paean to duty ("Duty! Wondrous thought, etc., etc.") by Immanuel Kant. Despite these distinctions between man and the higher animals, Darwin thought that "the difference in mind between man and the higher animals, great as it is, certainly is one of degree and not of kind" (494). Darwin objected to systems of classification that assigned a separate kingdom to the human race, for the difference in degree, however great, did not justify it. He stated: "If man had not been his own classifier, he would never have thought of founding a separate order for his own reception" (515).[61] Darwin

---

57. Alfred Russel Wallace, "Sir Charles Lyell on Geological Climates and the Origin of Species," *Quarterly Review* 126 (April 1869): 359–94.
58. Glacken notes that Darwin cited from Charles Bell, *The Hand: Its Mechanism and Vital Endowments as Evincing Design* (Philadelphia: Carey, Lea and Blanchard, 1833), 38.—Ed.
59. Darwin, *The Descent of Man*, chap. 3.
60. This is the opening sentence of chapter 4 of *The Descent of Man*.
61. The whole discussion is in chapter 6, pp. 512–51.

thought that there was a sharp break between early and primitive man and civilized man. Although he believed the same processes had been at work in human evolution as in other forms of life, the effects of natural selection in civilization were much diminished. The transition from early forms of human organization to civilization was not hard to explain: "With civilized nations, as far as an advanced standard of morality, and an increased number of fairly good men are concerned, natural selection apparently effects but little; though the fundamental social instincts were originally thus gained" (504). It was not Darwin but his enthusiastic supporters that were responsible for reimporting the struggle for existence (Malthusianism applied to nature as a whole) into the social world.

Darwin's analysis of civilization is most disappointing. It cannot be compared with that of Wallace, who saw clearly the miseries and the suffering of an industrial age dominated by laissez-faire capitalism. He was worried about the tendency of the weak members of society to propagate their kind. He saw advantages in the accumulation of capital, and disapproved of primogeniture, but civilization apparently favored "the better development of the body, by means of good food and the freedom from occasional hardships" (503). In Darwin's discussion there is no hint of compassion, no hint of child labor or slums; instead there are smug remarks that probably are more indicative of Darwin's remoteness from the sufferings of the laboring classes than of any innate lack of compassion. "Thus the reckless, degraded, and often vicious members of society, tend to increase at a quicker rate than the provident and generally virtuous member" (505). Apparently the Malthusian explanation of poverty blinded him to the evils of industrial society. Darwin's capacity for judging civilization was unequal to the realities of an industrial age, whose leaders complacently preached the doctrine of slow, gradual, and continuous progress under the leadership of applied science and technology. "The careless, squalid, uninspiring Irishman multiplies like rabbits: the frugal, foreseeing, self-respecting, ambitious Scot, stern in his morality, spiritual in his faith, sagacious and disciplined in his intelligence, passes his best years in struggles and in celibacy, marries late, and leaves her behind him."[62] This statement is in the most sordid tradition of the Western literature on natural character. Sum up a whole people of quickly, firmly, decisively, as if every member were the same, as if there were no squalid Scot, no sagacious Irishman!

In 1863, Thomas Henry Huxley published *Evidence as to Man's Place in Nature*, a treatise on the natural history of like apes, the relationships of man to the lower animals, and some fossil remains of man. With his customary insight, clarity, and forcefulness, Huxley saw the importance, four years after the publication of *The*

62. Darwin, *The Descent of Man*, 505.

Origin of Species, of considering man's place in nature from a perspective consistent with newly acquired knowledge and the theory of evolution.

> The question for mankind—the problem which underlies all others, and is more deeply interesting than any other—is the ascertainment of the place which Man occupies in nature and of his relations to the universe of things. Whence our race has come; what are the limits of our power over nature, and of nature's power over us; to what goal we are tending; are the problems which present themselves anew and with undiminished interest to every man born into the world.[63]

The passage sums up admirably the preoccupations of the time, for Huxley lived in a period when there was much interest in classification; nature and man's place in it, largely a product of the controversies over design, special creation, and evolution; man as a transformer of nature; environmental influences and progress.

Since Huxley accepted Darwinian theory, "man's place in nature" was not based on a concept that man was unique because he was a creation of God in His own image; neither was it different from the places occupied by the evolution of other forms of life. The gulf between humanity and the highest forms of animal life could be accounted for by evolutionary theory. It was impossible to erect "any cerebral barrier between man and the apes" (2:89); "the structural differences which separate man from the Gorilla and the Chimpanzee are not so great as those which separate the Gorilla from the lower apes" (2:96). He recognized, however, "the great gulf which intervenes between the lowest and the highest ape in intellectual power" (2:95). To those who were distressed and depressed over the apparent close relationship between the human race and the "brutal Chimpanzees and Gorillas," Huxley replied in essence that the lowly origin of man and his relationship with the chimpanzees and the gorillas was irrelevant.

> But it is not I who seek to base Man's dignity upon his great toe, or insinuate that we are lost if an Ape has a hippocampus minor. On the contrary, I have done my best to sweep away this vanity. I have endeavored to show that no absolute structural line of demarcation, wider that that between the animals which immediately succeed us in the scale, can be drawn between the animal world and ourselves; and I may add the expression of my belief that eh attempt to draw a psychical distinction is equally futile, and that even the highest faculties of feeling and of intellect begin to germinate in lower forms of life. At the same time no one is more strongly convinced than I am of the vastness of the gulf between civilized and the brutes; or is more certain that whether from them or not, he is assuredly not of them. (2:102)

63. T. H. Huxley, *Evidence as to Man's Place in Nature* (New York: D. Appleton, 1863), 2:53. Subsequent citations in the text are to this edition.

For Huxley, evolution had not destroyed but rather had created "the grandeur of the place" man occupies in the visible world (2:103).

In a course of six lectures to workingmen on the origin of species, Huxley, unlike Darwin, in an admirably clear way and without a trace of condescension saw in language the great dividing line between the human race and the higher primates.

> What is it that constitutes and makes man what he is? What is it but his power of language—that language giving him the means of recording his experience—making every generation somewhat wiser than its predecessor,—more in accordance with the established order of the universe? I say that this functional difference [the power of speech] is vast, unfathomable, and truly infinite in its consequences; and I say at the same time, that it may depend upon structural difference which shall be absolutely inappreciable to us with our present means of investigation.[64]

Like Darwin and Wallace, Huxley implied that an abstract term like "man" was unsatisfactory because of the differences between ancient, primitive, and civilized man; consequently, in his essay "The Struggle for Existence in Human Society" (1888), he also considered the question of whether the natural world offered any valid clues for understanding human society.[65] In approaching this subject, Huxley was closer to Wallace than to Darwin, for Huxley had close contacts with urban workers. He was interested in popular education, and he was aware of the human suffering in industrial society. These sides of Huxley, which seldom appear in histories of science or of nineteenth-century thought, are well portrayed in Cyril Bibb's *T. H. Huxley: Scientist, Humanist and Educator*.[66]

Following a typically cogent analysis of the shortcomings of physico-theology, Huxley made fun of extravagances in both the theological and the evolutionary positions.

> In the former we are told that this is a state of probation, and that the seeming injustices and immoralities of nature will be compensated by and by. But how this compensation is to be effected, in the case of the great majority of sentient things, is not clear. I apprehend that no one is seriously prepared to maintain that the ghosts of all the myriads of generations of herbivorous animals which lived during the millions of years of the earth's duration, before the appearance of man, and which have all that time been

---

64. Thomas H. Huxley, *On the Origin of Species, or, The Causes of the Phenomena of Organic Nature; A Course of Six Lectures to Working Men* (New York: D. Appleton, 1878), Lecture 6, pp. 148–49.

65. Thomas H. Huxley. "The Struggle for Existence in Human Society" (1888), in *Evolution and Ethics: And Other Essays* (New York: D. Appleton, 1894), vol. 9, 195–236. Subsequent citations in the text are to this edition.

66. Cyril Bibby, *T. H. Huxley: Scientist, Humanist and Educator* (London: Horizon Press, 1959).

> tormented and devoured by carnivores, are to be compensated by a perennial existence in clover; while the ghosts of carnivores ate to go to some kennel where there is neither a pan of water nor a bone with any meat on it. (198)

He was equally satirical about evolutionary teleology and the complacency it might engender.

> On the evolutionist side, on the other hand, we are told to take comfort from the reflection that the terrible struggle for existence tends to final good, and that the suffering of the ancestor is paid for by the increased perfection of the progeny. There would be something in this argument, if, in Chinese fashion, the present generation could pay its debts to its ancestors; otherwise it is not clear what compensation the *Eohippus* gets for his sorrows in the fact that, some millions of years afterwards, one of his descendants wins the Derby. (198–99)

Huxley met, in forthright fashion, the ancient dilemma of man and nature: whether man is or is not part of nature. An ambiguous controversy has continued in the environmental and ecological literature. These polarities have developed historically and cannot be considered apart from the history of ideas. It is important to recognize that it is impossible to understand the transformation of the environment by human agency if human beings are regarded, like beavers, as a part of nature. Understanding must rest on the distinction between human and other forms of actions in the environment. Huxley was not concerned with this question, but he saw some cogency in the ancient distinction between nature and art.

> In the strict sense of the word "nature," it denotes the sum of the phenomenal world, of that which has been, and is, and will be; and society, like art, is therefore a part of nature. But it is convenient to distinguish those parts of nature in which man plays the part of immediate cause, as something apart; and, therefore, society, like art, is usefully to be considered as distinct from nature. It is the more desirable, and even necessary, to make this distinction, since society differs from nature in having a definite moral object; whence it comes about that the course shaped by the ethical man—the member of society or citizen—necessarily runs counter to that which the non-ethical man—the primitive savage, or man as mere member of the animal kingdom—tends to adopt. (202–3)

This distinction, it would seem, constitutes an impassable barrier for any transfer of analogies from nature to human society, but Huxley did see the possibility of a struggle for existence in human society.

Huxley really abandoned the concept of abstract "man" in the sharp distinction he made between prehistoric and civilized man. Prehistoric men "were savages of a very low type." There was an authentic struggle for existence no different from the struggles among the animals. Beyond the limited and temporary relations of the family, "the Hobbesian war of each against all was the normal state of existence." In prehistoric human struggle, there was no place for praise or blame, no meaning to existence; planning and forethought were absent (203–4). "The history of civilization," by which he meant that of society, was, on the other hand, "the record of the attempts which the human race has made to escape from this position" (203–4). The ethical man, the product of civilization, "tries to escape from his place in the animal kingdom, founded on the free development of the principle of non-moral evolution, and to establish a kingdom of Man, governed upon the principle of moral evolution" (205).

To Huxley, the difficulty was that these strivings of ethical, civilized man did not abolish—indeed, perhaps hardly modified—"the deep seated organic impulses that impel the natural man to follow his non-moral course" (205). The key element in this crucial conflict was the tendency of population to multiply without limit; it is really the Malthusian principle of population, although Huxley does not say that it is. Genesis to him expressed a primordial urge of humanity. "It is notable that 'increase and multiply' is a commandment traditionally much older than the ten; and that it is, perhaps, the only one which has been spontaneously and *ex animo* obeyed by the great majority of the human race" (205–6). The danger is that this tendency of a population to multiply, characteristic of all life, will reestablish the struggle for existence in human society, "the mitigation or abolition of which was the chief end of social organization" (206).

When Huxley published this essay in 1888, Malthusian theory no longer commanded the widespread agreement it had enjoyed in the early decades of the nineteenth century. Its influence waned in the latter decades largely because of the increases in food production consequent on the opening up of the great chernozem soils to cultivation, a theme in Curtis Marbut's remarkable presidential address to the Association of American Geographers in 1925, which allayed the fears so forcefully aroused by Malthus.[67] John Maynard Keynes expressed a similar opinion in 1930: "After 1870 there was developed on a large scale an unprecedented situation, and the economic condition of Europe became during the next fifty years unstable and peculiar. The pressure of population on food, which had already been balanced by the accessibility of supplies from America, became for the first time in recorded history definitely reversed. As numbers increased, food was actually easier to se-

---

67. Curtis F. Marbut, "The Rise, Decline, and Revival of Malthusianism in Relation to Geography and Character of Soils," *Annals of the Association of American Geographers* 15 (1925): 1–29.

cure."[68] In the same work he also wrote, "Before the eighteenth century mankind entertained no false hopes. To lay the illusions which grew popular at that age's latter end, Malthus disclosed a Devil. For half a century all serious economical writings held that Devil in clear prospect. For the next half century he was chained up and out of sight. Now perhaps we have loosed him again."[69] What then caused Huxley to continue to accept the Malthusian doctrine in a period of decline and rejection of it? Perhaps there are two explanations, one contemporary, the other historical. The evils of industrial society and the sufferings of the urban poor were all known to Huxley. Beneath the glittering façade of apparent affluence, Huxley, like Wallace, saw the seamy side of civilization. Perhaps the principle of population could account for it. In addition, Huxley, like Malthus, thought that the tendency of the population to multiply had been a dynamic and directive force in the history of civilization. To Huxley, the conflict of rulers, the turbulence of the ruled, and wars were not "the causes of the decay of states and the foundering of old civilizations. . . . But beneath all this superficial turmoil lay the deep seated impulse given by unlimited multiplication."[70] Huxley was very close to Malthus in accepting this tendency (for the most part hidden) in determining the course of human history.

I am unaware of any suggestion in Huxley's work that he thought Darwinian theory had been inspired by the competitive struggle for existence in English industrial society of the nineteenth century. There was of course Malthusian theory, but it was really a biological theory that Malthus had applied to human society. Marx and Engels thought the theory of evolution was similarly inspired, but offered no proof beyond their assertions. If any scientists saw merit, Huxley and Wallace would have noted it because they were not remote from contemporary realties. In my opinion, Huxley thought the ideas of competition, the struggle for existence, and natural selection had their origin in the study of nature, not of society. It is a commonplace to suggest that abstract theories can be explained by religious, philosophical, technological, or ideological characteristics of the times. No doubt there is merit in this approach, but beyond flat assertions, the proof is difficult. Finally, there is the "Prolegomena" (1894) to *Evolution and Ethics*,[71] in which Huxley distinguished between the cosmic and the horticultural processes, the former having to do with nature, the latter with human activities.[72]

---

68. John Maynard Keynes, *The Economic Consequence of the Peace* (New York: Harcourt, Brace and Howe, 1920), 9.
69. Ibid., 10.
70. Ibid., 208.
71. Huxley, *Evolution and Ethics*.
72. Glacken notes, "I prefer to discuss it in the chapter on the study on the study and interpretations in the nineteenth-century of transformations of nature by human agency."—Ed.

Alfred Russell Wallace was the third of the evolutionist thinkers for whom man's place in nature was a vital element of the new evolutionary philosophy. Darwin and the many scientists and others who supported him believed that the human evolution had been a phase in the evolution of all life, and that physical, intellectual, moral man could be explained, if all the data were available, in the same manner as the evolution of other life forms was explained. Wallace departed from this firmly held belief, not with regard to the physical, but to the intellectual and moral nature of man. This departure was first apparent in a review of the tenth and entirely revised edition of Charles Lyell's *Principles of Geology* and the sixth edition of his *Elements of Geology*. Wallace remarked that Lyell carried Darwin's views "out to their legitimate results, and does not shrink from the logical necessity, of the derivation of man from the lower animals."[73] This agreement about the evolution of animal man did not, Wallace said, bar one from believing that the same process could not develop man's intellectual and moral nature. Thus Wallace introduced a new element into human evolution, and in consequence had a different reason for man's exalted place in nature and for the gulf between man and other higher forms of animal life: "Neither natural selection nor the more general theory of evolution can give any account whatever of the origin of sensational or conscious life.... But the moral and higher intellectual nature of man is as unique a phenomenon as was conscious life on its first appearance in the world, and the one is almost as difficult to conceive as originating by any law of evolution as the other."[74]

Wallace's argument that natural selection could not account for the development of the intellectual and moral qualities in man owed much to his firsthand experience with primitive peoples. His intimate knowledge of them comes out clearly in the ethnological passages of *The Malay Archipelago* (1872).[75] Wallace's attitudes toward them were not biased by the intense feelings of superiority of many believers in the idea of progress, who logically saw only a chasm between primitive and civilized man and to whom the word "savage" came quite naturally. Wallace used it too, but he had a deep sympathy for and understanding of primitive peoples. They were not the noble creatures of the eighteenth-century travelers and philosophers, nor were they howling savages, lacking a spark of feeling, pity, or justice: "The more I see of uncivilized people, the better I think of human nature on the whole, and the essential differences between civilized and savage man seem to disappear."[76]

---

73. Wallace, "Sir Charles Lyell...."
74. Ibid., 391.
75. Alfred Russell Wallace, *The Malay Archipelago: The Land of the Orangutan and the Bird of Paradise: a Narrative of Travel, with Studies of Man and Nature* (London: Macmillan, 1872).
76. Alfred Russel Wallace, *My Life: A Record of Events and Opinions*, 2 vols. (London: Chapman and Hall, 1905), 1:343.

Why should his attitudes toward primitive peoples encourage his belief that natural selection could not account for the intellectual and moral qualities of mankind? The reason is that Wallace believed primitive peoples were endowed with qualities that were unnecessary for their survival in the struggle for existence. Wallace was one of the few scientists of his time—perhaps the only one—who had acquired this unique knowledge of primitive peoples and their environments through prolonged association, a knowledge superior to that derived by missionaries interested in conversions, and who usually brushed aside primitive cultures as the work of the devil or at best the product of superstition. Huxley appreciated the unique quality of Wallace's experience. He wrote, "Once in a generation, a Wallace may be found physically, mentally, and morally qualified to wander unscathed through the tropical wilds of America and of Asia; to form magnificent collections as he wanders; and withal to think out sagaciously the conclusions suggested by his collections."[77] Although Wallace neither idealized primitive life nor advocated any return to nature, his observations of its unseen residues in civilization, the ignorance and superstition in civilized societies, did not allow him to draw polar contrasts between the intellect and morality of primitive and civilized people. Nor was he impressed by the difference in potential brain power between them. He used the comparative method, rather obliquely, to suggest a comparison between contemporary primitive peoples and prehistoric man.

> In the brain of the lowest savage, and, as far as we yet know, of the pre-historic races, we have an organ so little inferior in size and complexity to that of the highest types (such as the average European), that we must believe it capable, under a similar process of gradual development during the space of two or three thousand years, of producing equal average results. But the mental requirements of the lowest savages, such as the Australians or the Andaman islanders, are very little above those of many animals. The higher moral faculties and those of pure intellect and refined emotion, are useless to them, are rarely if ever manifested, and have no relation to their wants, desires or well-being. How then was an organ developed so far beyond the needs of its possessor? Natural selection could only have endowed the savage with a brain a little superior to that of an ape, whereas he actually possesses one but very little inferior to that of the average members of our learned societies. (391–92)

Wallace too was an uninhibited admirer of the human hand, which is well developed in civilized and primitive men, though the latter have no need for so fine an instrument,

---

77. T. H. Huxley, *Evidence as to Man's Place in Nature* (London: Williams and Norgate, 1863), 24.

and can no more fully utilize it than he could use without instruction a complete set of joiner's tools. But, stranger still, this marvelous instrument was foreshadowed and prepared in the Quadrumana; and any person, who will watch how one of these animals uses its hands, will at once perceive that it possesses an organ far beyond its needs. The separate fingers and the thumb are never fully utilized, and objects are grasped so clumsily, as to show that a much less specialized organ of prehension would have served its purpose quite as well; and if this be so, it could never have been produced through the agency of natural selection alone. (392)

Erect posture, "delicate yet expressive features," "the marvelous beauty and symmetry" of the human form are unique possessions of man, unequaled among the higher animals.

> Those who have lived much among savages know that even the lowest races of mankind, if healthy and well fed, exhibit the human form in its complete symmetry and perfection. They all have the soft smooth skin absolutely free from any hairy covering on the dorsal line, where all other mammalia from the Marsupials up to the Anthropoid apes have it most densely and strongly developed. What use can we conceive to have been derived from this exquisite beauty and symmetry and this smooth bare skin, both so very widely removed from his nearest allies? (392)

Human speech could hardly have been of much use to the lowest savages. Such skills could not have been developed by natural selection. Among the lowest savages "the capacity of uttering a variety of distinct articulate sounds, and applying to them an almost infinite amount of modulation and inflection, is not in any way inferior to that of the higher races. An instrument has been developed in advance of the needs of its possessor" (393).

Wallace's solution to the problem dismayed scientists and evolutionists in particular. In essence, he denied that the theory of evolution expressed "the whole truth regarding the origin of man." He granted that the same laws of organic development operated in the origin of man as in the rest of life, but that there seemed to be "evidence of a Power which has guided the action of those laws in definite directions and for special ends" (393). In support of this argument, Wallace drew, in a breathtaking manner, an analogy from human experience. It is very interesting because man is cast in the role not of an observer alone, but of a participant and an agent in directing natural processes: "Man himself guides and modifies nature for special ends. The laws of evolution alone would perhaps never have produced a grain so well adapted to his uses as wheat; such fruits as the seedless banana, and the bread-fruit; such animals as the Guernsey milch-cow, or the London dray-horse" (393). Man has been able to guide the laws of variation, multiplication, and survival for his own purposes. We should not shut our eyes to evidence that "'an Overruling Intelligence' has watched over the action of those laws, so directing variations and so determining their accumulation, as finally to produce an organization sufficiently perfect to admit of, and even to aid in, the indefinite advancement of our mental and moral nature" (394). This reasoning would not have been possible if Wallace had held a different conception of the nature of man, who is conceived as an intelligent, purposeful agent, capable of modifications, by domestication, of nature, and suggesting by analogy a similar but infinitely more exalted role of an "Overruling Intelligence." It is the old idea of an artisan deity in evolutionary dress. It would seem, therefore, that his argument as to the inadequacy of evolutionary theory to account for human evolution was based on the characteristics of primitive peoples and on the power of the human race to transform nature.

It is quite evident that Wallace's experience with primitive peoples was a decisive factor in formulating his ideas concerning the inadequacy of natural selection to develop the intellectual and moral qualities of man. The culture of contemporary primitive peoples also became a guide to the possible nature of ancient man. Like many nineteenth- and twentieth-century proponents of the comparative method, Wallace did not consider the possibility that such a comparison might be in error; that contemporary primitive peoples, like civilized peoples, have a history; that they had been in contact with other primitive peoples or peoples of higher civilizations; and that their characteristics and values, therefore, might not be indicative of those of ancient man.

One of the distinguishing characteristics of civilized man as compared with "savages," in Wallace's opinion, was the ability to formulate abstract ideas. The languages of savages have no words for abstract "conceptions." The savage lacks foresight; he is unable "to combine, or to compare, or to reason" on any general subject that does not immediately appeal to his senses. He also lacks

> those conceptions of the infinite, of the good, of the sublime and beautiful, which are so largely developed in civilized man. Any considerable development of these would, in fact, be useless or even hurtful to him, since they would to some extent interfere with the supremacy of those perceptive and animal faculties on which his very existence often depends, in the severe struggle he has to carry on against nature and his fellow-man. Yet the rudiments of all these powers and feelings undoubtedly exist in him, since one or other of them frequently manifest themselves in exceptional cases, or when some special circumstances call them forth.[78]

78. Alfred Russel Wallace, "The Limits of Natural Selection as Applied to Man," in Wallace, *Contributions to the Theory of Natural Selection* (London: Macmillan, 1870), 340–41.

An example of this last point were the Santals of India, who, he claimed were "remarkable for as pure love of truth as the most moral among civilized men" (341). He also remarked that the Hindus and the Polynesians had a high artistic feeling, the first traces of which were clearly visible in "the rude drawings of the Paleolithic men who were contemporaries in France of the Reindeer and the Mammoth" (341).

Wallace could see how abstract justice and benevolence might have been acquired through natural selection because of their survival value to the tribe, but natural selection could not account for another class of faculties unconcerned with fellow men, such as the capacity to form ideal conceptions of space and time, of eternity and infinity; the capacity for intense artistic feelings of pleasure, in form, color, and composition; and the capacity for those abstract notions of forms and number that render geometry and arithmetic possible. How were all or any of these faculties first developed, if they could have been of no possible use to man in his early stages of barbarism? How could "natural selection," or survival of the fittest in the struggle for existence, ever have favored the development of mental powers so entirely removed from the material needs of savage men, and which even now, with our comparatively high civilization, are, in their farthest developments, in advance of the age, and appear to relate more to the future of the race than to its actual status (351–52)? Wallace also rejected the argument that a moral sense and a conscience developed in savage men because such traits were useful to the tribe possessing them; this explanation did not account for "the peculiar *sanctity* attached to actions which each tribe considers right and moral, as contrasted with the very different feelings with which they regard what is merely *useful.*" The utilitarian hypothesis, he argued, seems "inadequate to account for the development of the moral sense" (352).

No stranger to hornet's nests, and in his characteristically courteous and forthright manner, Wallace clarified what he was driving at.

> I can only explain this misconception by the incapacity of the modern cultivated mind to realize the existence of any higher intelligence between itself and Deity. Angels and archangels, spirits and demons, have been so long banished from our belief as to have become actually unthinkable as actual existences, and nothing in modern philosophy takes their place. Yet the grand law of "continuity," the last outcome of modern science, which seems absolute throughout the realms of matter, force, and mind, so far as we can explore them, cannot surely fail to be true beyond the narrow sphere of our vision, and leave an infinite chasm between man and the Great Mind of the universe. Such a supposition seems to me in the highest degree improbable. (372, note A)

Without passing on the merits of his argument, it must be admitted that this is a brilliant sketch of a crucial development in modern Western thought in which archangels, seraphs and seraphims, demons, devils, and other angelic and not so angelic creatures were unceremoniously thrown overboard, leaving vast areas unguarded, uncared for, and unlived in.

Darwin was distressed at this turn in Wallace's thought. In his letter of April 14, 1869, Darwin complimented Wallace enthusiastically on the *Quarterly Review* article. The Darwin-Wallace correspondence is touching to read because of the friendly, generous, sensitive appreciation they had for one another. He wrote, "Your exposition of Natural Selection seems to me inimitably good; there never lived a better expounder than you. . . . Altogether I took at your article as appearing in the 'Quarterly' as an immense triumph for our cause." Of Wallace's remarks on man, however, Darwin said, "If you had not told me I should have thought that they had been added by some one else. As you expected, I differ grievously from you, and I am very sorry for it. I can see no necessity for calling in an additional and proximate cause in regard to man."[79] Wallace, however, did not change the opinion he had expressed in 1869, and the summary of his views in his journal, written in 1905, was if anything more incisive, succinct, and positive than previous statements.[80]

Despite Darwin's dismay that Wallace entertained such thoughts, and because of his own intellectual and religious history, he was in a good position to understand them. In 1876, Darwin wrote in his journal a rather complete statement of his religious views. The most striking of them is the idea, prominent in the eighteenth century and in the Romantic movement of the nineteenth century, of the influence of sublime scenery on religious belief.

> In my Journal I wrote that whilst standing in the midst of the grandeur of a Brazilian forest, "is it not possible to give an adequate idea of the higher feelings of wonder, admiration, and devotion, which fill and elevate the mind." I well remember my conviction that there is more in man than the mere breath of his body. But now the grandest scenes would not cause any such convictions and feelings to rise in my mind.[81]

To him, inward convictions and feelings were not evidence of what really exists.

> The state of mind which grand scenes formerly excited in me, and which was intimately connected with a belief in God, did not essentially differ from that which is often called the sense of sublimity; and however difficult it may be to explain the gen-

---

79. F. Darwin, ed., *Life and Letters*, 2:296–97.
80. Wallace, *My Life*, 1:427–28.
81. Charles Darwin, *The Autobiography of Charles Darwin 1809–1882: With the Original Omissions Restored*, edited and with appendix and notes by his granddaughter, Nora Barlow (London: Collins, 1958). [Also in F. Darwin, ed., *Life and Letters*, 1:281.—Ed.]

esis of this sense, it can hardly be advanced as an argument for the existence of God, any more than the powerful though vague and similar feelings excited by music.[82]

Darwin was talking about sublime scenery, but in a broader sense he was rejecting one of the oldest ideas in Western civilization, with roots in both the classical and the Judeo-Christian worlds, that the beauties and harmonies of nature are evidence of the existence of God. He wrote,

> This follows from the extreme difficulty or rather impossibility of conceiving this immense and wonderful universe, including man with his capacity of looking far backwards and far into futurity, as the result of blind chance or necessity. When thus reflecting I feel compelled to look to a First Cause having an intelligent mind in some degree analogous to that of man; and I deserve to be called a Theist. This conclusion was strong in my mind about the time, as far as I can remember, when I wrote the 'Origin of Species;' and it is since that time that it has very gradually, with many fluctuations, become weaker. But then arises the doubt; can the mind of man, which has, as I fully believe, been developed from a mind as low as that possessed by the lowest animals, be trusted when it draws such grand conclusions? . . . I cannot pretend to throw the least light on such abstruse problems. The mystery of the beginning of all things is insoluble by us; and I for one must be content to remain an Agnostic.[83]

This passage, along with that quoted from Wallace, reveals the continuing strength and force of the ancient artisan analogy in nineteenth-century thought—the conception of God as a divine workman, fashioning, like an artisan, all creation. To all appearances, Darwin was very uncertain in his interpretation of nature, considered both cosmically and terrestrially. Design in the cosmos, apparently, was a constantly recurring thought, while at the same time he rejected design and final causes in interpreting interrelationships in nature on Earth.

The thoughts of Darwin, Huxley, and Wallace on the place of man in nature are the culmination of speculations, opinions, and theories that began in ancient times. Historically in Western civilization, concepts of the nature of the natural world on Earth have, quite understandably, been inseparable from those of the place of the human race in nature for the obvious reason that human beings live on an Earth clothed in plants and inhabited by animals, birds, and insects. Such circumstances have produced bodies of thought concerned with the differences between human and other forms of life; with the clear superiority of human over other forms of life in the sense that human beings could select favored plants, and impose their

82. F. Darwin, ed., *Life and Letters*, 1:281.
83. Ibid., 1:282.

will on dogs, oxen, horses; and with a unique relationship between the human race and God or the gods. For centuries, the accepted view among earthly beings was that the human race occupied the highest place in the scale of being created permanently from the beginning by God. The representative views of Darwin and Huxley—and of Wallace also, despite his reservations—replaced them. The place of man in nature was as exalted as that envisaged by the natural theologians. The difference was in the manner in which this high place had been attained.

Since their time, a shift in emphasis has occurred. The idea of man's place in nature still remains a powerful idea in Western thought. When we use the expression today, we do not mean the place of the human race in the evolutionary scheme of things but in an ecological sense, the relationship of human cultures to a physical environment conceived of as an ecosystem. The persistence of the idea of man's place in nature owes much to the ecological aspects of Darwinian theory. The shift was from man's place in the evolutionary scale to his place in the web of life. Only a few scientists and other informed observers saw this development, so immersed were they in natural selection, variation, evolution, in controversies often within science itself rather than between science and theology. J. Arthur Thomson in 1909 recognized the ecological and stressed it. Among the services Darwin rendered to the theory of organic evolution, Thomson said, "as an epoch-marking contribution, not only to Aetiology but to Natural History in the widest sense, we rank the picture which Darwin gave to the world of the web of life, that is to say, of the inter-relations and linkages in Nature."[84] Perhaps Thomson gives too much credit to Darwin, Buffon, Lamarck, and Humboldt, and many natural theologians expressed similar ideas, but it is true that ecological studies in the latter part of the nineteenth and the early part of the twentieth centuries often show Darwin's influence, although others came out of the earlier history of plant geography.

To place man in nature was a central theme of Marsh's *Man and Nature*; it was based on ideas of a balance in nature, and of human intrusions onto this balance. The middle and latter parts of the nineteenth century mark the beginnings of an intensified interest in man's place in nature from an ecological point of view rather than as a highly conspicuous end product of the evolutionary process. Immediately following the discussion of northern influences on the south, Wallace, in an abrupt transition, applied in a dismaying manner the struggle for existence to human society. This struggle leads

---

84. J. Arthur Thomson, "Darwin's Predecessors," in *Darwin and Modern Science: Essays in Commemoration of the Centenary of the Birth of Charles Darwin and of the Fiftieth Anniversary of the Publication of The Origin of Species*, ed. A. C. Stewart (Cambridge: Cambridge University Press, 1909), 3–17, at 4.

to the inevitable extinction of all those low and mentally undeveloped populations with which Europeans come into contact. The Red Indian in North America, and in Brazil; the Tasmanian, Australian, and New Zealander in the southern hemisphere, die out, not from any one special cause, but from the inevitable effects of an unequal mental and physical struggle. The intellectual and moral, as well as the physical, qualities of the European are superior; the same powers and capacities which have made him rise in a few centuries from the condition of the wandering savage with a scanty and stationary population to his present state of culture and advancement, with a greater average longevity, a greater average strength, and a capacity of more rapid increase,— enable him when in contact with the savage man, to conquer in the struggle for existence, and to increase at his expense, just as the better adapted, increase at the expense of the less adapted varieties in the animal and vegetable kingdoms,—just as the weeds of Europe overrun North America and Australia, extinguishing native productions by the inherent vigor of their organization, and by their greater capacity for existence and multiplication.[85]

It is hard to make sense of this passage. Is the comparison with weeds bitter and ironic? No such dithyrambic assessment of European civilization occurs elsewhere in Wallace's writings. Why did Wallace, like so many of his contemporaries, substitute for the historical record monistic explanations such as the struggle for existence as determinants in the extermination of so-called primitive peoples? The passage is puzzling because it is so out of keeping with the closing pages of *The Malay Archipelago*, and with his sympathy for primitive peoples and his criticisms of contemporary industrial civilization. Wallace's analysis seems to be a confused combination of historical and evolutionary thought. In one part of the passage cited, he relied on highly generalized history, and in the other on the struggle for existence imported from the biological world. Is it appropriate, however, to speak about the successes of civilized over primitive peoples as an example of the struggle for existence? The extermination of primitive peoples in the New World after the age of discovery, in the islands of the Pacific, and in many other parts of the world during the nineteenth century (and continuing on to the present) was not the result of a struggle between equally equipped antagonists (as Wallace notes), and the European forces and the [primitive peoples] were not involved in a struggle for their existence.[86] The proper way of studying the relationships between civilized and primitive peoples is to study history, not to import ideas from evolutionary theory. Teggart saw this truth long ago. Commenting on the pitfalls of using expres-

---

85. Alfred Russel Wallace, "The Development of Human Races under the Law of Natural Selection," in Wallace, *Contributions to the Theory of Natural Selection*, 302–31, at 319.
86. There is a brief lacuna in the text; "primitive peoples" is our interpolation.—Ed.

sions like "struggle for existence," "survival of the fittest," and "natural selection," he said, "The student of the evolution represented in the facts of human history must, therefore, be prepared to take upon himself the burden of an independent investigation; he cannot hope to adopt ready-made the formulae which have proved useful in other subjects; and he will turn to Darwin simply to observe the method which he employed."[87] Commenting further on Darwin's recognition of the dangers in using these terms, Teggart added, "The difficulty, which presents itself even in biology, is, however, greatly increased when these words, freighted with new meanings, are carried back again into the discussion of social problems."[88]

The tone of the closing pages of *The Malay Archipelago* is very different. The essay on the development on the human race was first published, revised, and republished in 1871. *The Malay Archipelago* was first published in 1869; the preface to the tenth edition is dated 1890. In those works, Wallace discussed what civilized people can learn from "the savage man." Without idealizing them, without hints of a kinship with the "noble savage" of the eighteenth century, without any expressed sympathy with primitivism or a return to nature, Wallace simply said that "it is very remarkable that among people in a very low stage of civilization we find some approach" to the perfect social state that is aspired to by many thinkers of the civilized world:

> I have lived with communities of savages in South America and in the East, who have no laws or law courts but the public opinion of the village freely expressed. Each man scrupulously respects the rights of his fellow, and any infraction of those rights rarely or never takes place. In such a community, all are nearly equal. There are none of those wide distinctions, of education and ignorance, wealth and poverty, master and servant, which are the product of our civilization; there is none of that wide-spread division of labor, which, while it increases wealth, produces also conflicting interests; there is not that severe competition and struggle for existence, or for wealth, which the dense population of civilized countries inevitably creates.[89]

In effect, he argued that "we" have progressed intellectually, but not morally, beyond the savage:

> During the last century, and especially in the last thirty years, our intellectual and material advancement has been too quickly achieved for us to reap the full benefit of it. Our mastery over the forces of nature has led to a rapid growth of population, and

---

87. Frederick J. Teggart, *Prolegomena to History: The Relation of History to Literature, Philosophy, and Science*, vol. 4 in History series (Berkeley: University of California Press, 1916), 225.
88. Ibid., 225n38.
89. Wallace, *Malay Archipelago*, 456.

a vast accumulation of wealth; but these have brought with them such an amount of poverty and crime, and have fostered the growth of so much sordid feeling and so many fierce passions, that is may well be questioned whether the mental and moral status of our population has not on the average been lowered, and whether the evil has not overbalanced the good.[90]

This was a far cry from the paeans to progress frequently expressed in the nineteenth century, a far cry from the triumphs of the Crystal Palace.[91] With an insight that we in the twentieth century with its slums, its jungles, its bidonvilles in large cities the world over have scarcely achieved, Wallace said, "We should now clearly recognize the fact, that the wealth and knowledge and culture of the few, do not constitute civilization, and do not of themselves advance us towards the 'perfect social state' state."[92] The concluding pages, in which he condemned modern civilization as but a form of barbarism, "procured me the acquaintance of John Stuart Mill," who wrote Wallace from Avignon on May 19, 1870, proposing the organization of a Land Tenure Reform Association and asking Wallace to become a member of the General Committee.[93]

---

90. Ibid., 456–57.
91. John Bagnell Bury, *The Idea of Progress An Inquiry into Its Origin and Growth* (London: Macmillan, 1920). See esp. chap. 18, on the Exhibition of 1851. [The Crystal Palace refers to a building erected in Hyde Park in London to house the Great Exhibition of 1851. The exhibition was a celebration of the Industrial Revolution.—Ed.]
92. Ibid., 457.
93. Wallace, *My Life*, 2:235.

IN THE WAKE OF MARSH

This essay, which consolidates portions of Glacken's handwritten manuscript, "Man and Nature: Selected Essays," turns to some well-known and relatively unknown nineteenth- and twentieth-century writers, and in doing so addresses two of the big ideological ideas in modern environmentalist thought, that of anthropogenic change and that of carrying capacity. The essay is organized in five parts. It begins with George Perkins Marsh, and, while addressing a host of themes in his classic opus, *Man and Nature,* including the impact of deforestation, dwells on Marsh's concern with "the subjugation of the entire organic and inorganic world to human control and human use."

The discussion of all the other authors considers their varied attempts, over several decades, to refine the concept of carrying capacity. The first author discussed is Friedrich Ratzel, famous (or infamous) for proposing the concept of *Lebensraum.* Glacken discusses Ratzel's ideas about the distribution of the habitable and natural worlds, and the capacity of Europe to expand and of its populations to migrate to other parts of the world. Though Glacken understates any criticism of Ratzel's politics and ideologies, he remarks, "His is not the habitable world of Marsh being torn away by human agency; it is a world filling up by the struggle for space, under the leadership of European culture, itself a product of that struggle." There follows a discussion of the geographer E. G. Ravenstein and his methods of estimating the world's population, and especially of calculating just how many people the Earth can support. This theme is renewed with all the other writers discussed in this essay, of whom the next is the geographer Carl Bellod, who was concerned additionally with the problem of the exhaustion of fuels in relation to the habitability of the temperate and tropical worlds and with soil quality, which for him was a significant variable. After Bellod, Glacken discusses the American geneticist Edward M. East, who pursued similar research questions but arrived at strong Malthusian conclusions, which led him to advocate worldwide birth control. Finally, there is an account of the work of the geographer Albrecht Penck and others who came in his wake, who also were concerned with similar questions.

Although Glacken's method—which consists of summaries and brief commentaries—does not allow him to explore the wider contexts and consequences of these writings, this essay provides a useful background for understanding the origins and persistence of two of modern environmentalism's most persistent and controversial concepts. They also make the reader pause, and think about the ideological implications of environmental theories, and the way they have been used and abused in history. Some of the protagonists in this chapter, for example, provided significant intellectual fodder for the Nazi canon, even if they found themselves on the wrong side, as did, for example, General Karl Haushofer, founder of the journal *Zeitschrift für Geopolitik*, to which Fischer contributed. It is also quite striking, from the vantage point of the twenty-first century, to see how widespread colonialism and imperialism were in the early days of environmentalist thought. Although Glacken does not pursue these wider issues in this essay, they could well be profitable subjects for future scholars to pursue.

George Perkins Marsh

Until shortly after the middle of the nineteenth century, writers considering the relation of man to his environment had dealt primarily either with the influence of the physical environment on man or with obstacles in society that prevented man's complete and happy conquest of the Earth. There was no hint, at least on a global scale, that as a result of these efforts, wide-ranging and destructive forces had been set in motion by man, and one could no longer complacently view the world as a harmonious whole or assume that modifications in the environment coming from cultivation, public works, or deforestation were in their nature improvements to the physical environments of man.

The man responsible for this change was an American, and we must view his performance as an exceptionally interesting fusion of many ideas. In 1864 George Perkins Marsh published *Man and Nature; or, Physical Geography as Modified by Human Actions*.[1] The fusion of ideas I mentioned came from three chief sources: first, the exceptionally wide range of interests of Marsh himself, enriched by his facility in many European languages; second, Marsh's opportunity for travel in the

---

1. George Perkins Marsh, *Man and Nature, or, Physical Geography as Modified by Human Action* (New York: C. Scribner, 1864). Subsequent citations in the text are to this edition.

Near East and Western Europe, areas of long human settlement; and third, the opportunity to witness, study, and follow the great westward expansion taking place in the United States. Thus there was a great contrast presented to a sensitive and deep intellect, the contrast between unsettled lands being opened up and lands that had been settled for millennia.

George Perkins Marsh (1801–82) was born in Vermont, and in early life divided his time among the law, business, and scholarly pursuits. He also mastered several languages, including French, Spanish, Portuguese, German, Italian, and the Scandinavian languages. He cultivated an interest in the etymology of the English language and assisted James Murray in the early work on the *Oxford English Dictionary*.[2] He also studied geology and immersed himself in the natural sciences, following, it would seem, a tradition set by Thoreau. In 1825 he was admitted to the bar, and in 1834 elected to Congress as a Whig. During his career in Congress he supported a high tariff, but was opposed to slavery and the Mexican War.[3] President Taylor in 1849 appointed him minister to Turkey, where he served til his recall in 1854. It was this service in Turkey that gave him the opportunity to travel widely throughout the old Turkish Empire, and especially in the Near East, in particular Egypt. On his return to the United States he became railroad commissioner for the state of Vermont, and his interests in both railroads and the law appear frequently in *Man and Nature*. These practical pursuits did not prevent him from delivering the Lowell Lectures in English philology and etymology in the winter of 1860–61.[4] His second opportunity for wider acquaintance with the Near East and Western Europe occurred in 1860 when President Lincoln appointed him first American minister to the Kingdom of Italy, whose unification was secured in that year, though not without further struggle. Marsh remained in this post for the rest of his life, a period of twenty-one years.

It is easy to see how an inquiring mind with Marsh's background and in the presence of these contrasts would question the easygoing, one might almost say conventional, generalizations current regarding the influence of climate, topog-

---

2. Sir James Augustus Henry Murray (1837–1915) was a Scottish lexicographer and philologist who was the editor of the *Oxford English Dictionary* from 1879 until his death.—Ed.

3. The Mexican War, also known as the Mexican-American War, was a conflict between the United States and the Centralist Republic of Mexico from 1846 to 1848, in the wake of the 1845 U.S. annexation of Texas.—Ed.

4. During the winter of 1860–61 Marsh gave a series of lectures titled "The Origin and History of the English Language and of the Early Literature That It Embodies" at the Lowell Institute in Boston (New York: Charles Scribner, 1862; rev. ed., 1881).These lectures were historical, chronologically tracing the history of the language from the first lecture, "Origin and Composition of the Anglo-Saxon People and Their Language," to "The English Language and Literature during the Reign of Elizabeth."—Ed.

raphy, location, and all the other ideas that had changed and been refuted since the times of Hippocrates, Herodotus, Polybius, and Strabo. The case, he felt, was just as strong for man's modification of his environment, rather than the other way around. These ideas are already apparent, but without the pessimism, in his efforts to persuade the American government to introduce the camel into the United States, especially in the arid and desert regions west of the Mississippi. In 1854 he had delivered a lecture on this subject at the Smithsonian Institution that was later expanded into a book, *The camel, his organization, habits and uses, considered with reference to his introduction into the United States.*[5] He had studied the camel carefully and read all the literature on this animal, including the long essay on its distribution that Ritter had prepared in the thirteenth volume of his *Erdkunde.*[6] He spoke of the sympathy that the "present secretary of war" had for his project, acknowledging in his book that Jefferson Davis had allowed him to read the official army correspondence on the subject of camels.[7]

In his book on the camel, Marsh announced some of the ideas that were to be much more fully developed in his later works. He started out with the command from Genesis 2, but applied it in a different way than the older writers against Malthus and for population increase had. The command was to go forth and multiply, subdue the Earth, and secure dominion over terrestrial features. It "thus predicted and prescribed the subjugation of the entire organic and inorganic world to human control and human use."[8] In this mankind had not yet succeeded, for there were millions of leagues of the Earth not yet subdued and not yet inhabited. Thus far, human successes had been mostly in gaining domination over the inorganic world, such as by the exploitation of minerals, and developing some control over natural forces. Despite the domestication of plants, few of the available world plants had been used. Marsh, however, was interested in the converse question, "the importance of human life as a transforming power," which, he suggested, "is perhaps,

---

5. George Perkins Marsh, *The camel, his organization, habits and uses, considered with reference to his introduction into the United States* (Boston: Gould and Lincoln, 1856). See also David Lowenthal, *George Perkins Marsh: Prophet of Conservation* (Seattle: University of Washington Press, 2009), 174.—Ed.

6. Carl Ritter, *Die Erdkunde im Verhältniss zur Natur und zur Geschichte des Menschen* [Geography in relation to nature and the history of mankind], 1816–1859 (Berlin: G. Reimer, 1862). —Ed.

7. Glacken notes, without providing sources, that "The second session of the Thirty-third Congress had appropriated $30,000 for importing camels for army transportation and other purposes and army officers sent abroad to purchase that had actually landed 32 on the coast of Texas."—Ed.

8. Marsh, *The Camel,* introduction.

more clearly demonstrable in the influence man has thus exerted upon superficial geography than in any other result of his material effort" (iv–v).[9]

Marsh's overall view on this question is expressed in the introductory chapter of his work. He took as his main example the Roman Empire at its greatest extent. The Roman example was decisive in his mind because no other place on Earth had enjoyed such a combination of factors suited to human habitation. He began by writing that "the provinces bordering on the principal and secondary basins of the Mediterranean enjoyed a healthfulness and an equality of climate, a fertility of soil, a variety of vegetable and mineral products, and natural facilities for the transportation and distribution of exchangeable commodities, which have not been possessed in equal degree by any territory of like extent in the Old World or the New" (2). He went on to observe that the provinces of this empire, such as northern Africa, the greater Arabian peninsula, Syria, Mesopotamia, Armenia, many provinces of Asia Minor, Greece, Sicily, and parts of Greece and Spain, so fertile and hospitable in ancient times, were now so infertile and exhausted as to be of little use of civilized man. If the wasted soils of Persia and the remaining eastern lands are added to this, he argued, "a territory larger than all Europe . . . has been entirely withdrawn from human use, or, at best, is thinly inhabited by tribes so few in numbers, too poor in superfluous products, and too little advanced in culture and the social arts, to contribute anything to the general moral or material interests of the great commonwealth of man" (5).

What were the causes of all this? Marsh did not rule out geology or meteorology, but in his opinion the main cause was human activity. Roman despotism, the feudalism that followed it, and the corruption of the Christian Church: the causes could be summed up as violations of the laws of nature, war, and civil and ecclesiastical misrule. A realization of these developments came later, Marsh wrote,

> and it is but recently that, even in the most populous parts of Europe, public attention has been half awakened to the necessity of restoring the disturbed harmonies of nature, whose well-balanced influences are so propitious to all her organic offspring, of repaying to our great mother the debt which the prodigality and thriftlessness of former generations have imposed upon the successors—thus fulfilling the command of religion and practical wisdom, to use this world as not abusing it. (12)

This was the reason Marsh thought geographers who were tracing the influence of geography on man were on the wrong track. The causes of changes in the Earth's surface are complex; meteorology, and especially the mechanism of climatic change, are little understood; there is a general paucity of facts, but, qualitatively

---

9. Marsh, *Man and Nature*, preface.

considered, human activity has been a very great influence. This has not all been bad, but the weight of influence has been on the destructive side.

There were circumstances in this age, Marsh argued, that gave this subject a practical and immediate interest. One was the need to provide new homes for a European population, "which is increasing more rapidly than its means of subsistence, new physical comforts for classes of the people that have now become too much enlightened and have imbibed too much culture to submit to a larger deprivation of the share of the material enjoyments which the privileged ranks have hitherto monopolized" (27). And after the "new hives for the emigrant swarms"—in Australia, the oceanic islands, South and Central Africa, on the shores of the Mediterranean, and on the prairies and forests of America—have been filled, it will be necessary to increase and make production more efficient for those in Europe who remain behind (27). And a third reason lay in the altitude of the unsettled regimes.

The idea of usufruct, not consumption or waste, might be central either to restoring old lands or to settling new ones. Nature, Marsh wrote, "has left it within the power of man [irreparably] to [diverge] to the combinations of inorganic matter and organic life, which through the night of aeons she had been proportioning and balancing, to prepare the earth for his habitation, when, in the fullness of time, his Creator should call him forth to enter into its possession" (36). Intentional changes were important enough, but it was the unsought and unexpected modifications that were the greatest influence. At a time when Darwin had already published *The Origin of Species,* and almost thirty years before Huxley told the world that nature was a gladiatorial show, Marsh was pointing out the importance of nature in the face of the higher being, man: "The fact that, of all organic beings, man alone is to be regarded as essentially a destructive power, and that he wields energies to resist which, nature—that nature whom all material life and all inorganic substances obey—is wholly important, tends to prove, that, though living in physical nature, he is not of her, that he is of more exalted parentage, and belongs to a higher order of existence than those born of her womb and submissive to her dictates" (36–37). Animal destructiveness was balanced by compensations; their populations were controlled by the law of supply and demand. With man, progress consisted in combating nature through art and many interferences; for man had gone beyond what was necessary. Transformations were a part of advancing civilization; primitive peoples made little permanent change. "The earth is fast becoming an unfit home for its noblest inhabitant, and another era of equal human crime, and human improvidence, and of like duration with that through which traces of that crime and that improvidence, extend, would reduce it to such a condition of impoverished productiveness, of shattered surface, of climatic excess, as to threaten the depravation, barbarism, and perhaps even extinction of the species" (44).

Marsh thought there was hope. Some forests had been replanted, and there had been many other improvements of a partial character: "It is, on the one hand, rash and unphilosophical to attempt to set limits to the ultimate power of man over inorganic nature, and it is unprofitable, on the other to speculate on what may be accomplished by the discovery of now unknown and unimagined forces, or even by the invention of new arts and processes" (45–46). The restoration of the old countries required political and moral revolutions and a knowledge of climates and soils. With few exceptions, these countries could be expected to deteriorate even more, "and in the meantime, the American continent, Southern Africa, Australia, and the smaller oceanic islands, will be almost the only theatres where man is engaged, on a great scale, in transforming the face of nature" (48).

No doubt each modern reader will come away from Marsh with a different viewpoint. To some, his words seem like a prophetic warning; to others his vision will seem overdrawn and unfilled. The South Africans have a problem of soil erosion that is the preoccupation of the area's best workers in the fields, to say nothing of one of its prominent novelists.[10] The Australians, largely because of the early efforts of Taylor, have been attempting to show that their subcontinent cannot be the habitation of millions.[11] And the American experience is written in the legislation of the nineteenth and twentieth centuries on soil conservation and erosion. But in this new geography of 1864, there is an impressive awareness of the influence of historical events on Earth—not a general awareness but a specific one. The materials in large part came not only from observation but also from the collection of writings on local areas. The greatest transformations date from around the beginning of the Christian era, and there is strong emphasis on the cumulative effect of these modifications. Marsh's book vividly depicts the tremendous impact that deforestation had on the thinking of nineteenth-century writers concerned with these subjects. It is an indictment of the nineteenth century only in the sense that it has added to this accumulation and that its restorations were only partial. The book purports to show the progressive deterioration of the habitable Earth, but its material fails in this effort, for the illustrations are largely from Western Europe. Despite frequent mention of soil erosion, there is little treatment of soils as such,

---

10. Glacken does not provide a reference, but see Christopher Heywood, *A History of South African Literature* (Cambridge: Cambridge University Press, 2004), esp. chap. 5, for a description of the key authors of this period and their contexts.—Ed.

11. Glacken was very likely referring to Thomas Griffith Taylor (1880–1963). For a brief biographical account, including his controversial environmental determinist statements, see J. M. Powell, "Taylor, Thomas Griffith (1880–1963)," in *Australian Dictionary of Biography* (Canberra: Australian National University, National Centre of Biography, 1990), http://adb.anu.edu.au/biography/taylor-thomas-griffith-8765/text15363 (accessed August 28, 2015).—Ed.

and Marsh seems to rely largely on Jean-Baptiste Boussingault.[12] On Liebig, Marsh has little to say, except to take issue with him. Marsh's great example was deforestation, and the exhaustion of natural resources, and even soil erosion as a subject in itself, are secondary to this main theme. But even Marsh wrote before the great industrialization of the nineteenth century, and historians conventionally date the wide-scale industrialization of the United States from the Civil War. Marsh wrote at a time when kerosene was used chiefly for lamps and before the vulcanization of rubber, in 1839, had late made a great industry of bicycle tires. However, awareness of the "habitable world" was not approached by any previous writer, and it is only the global studies—conducted, say, since World War I—of soils, ecology, and trade that have permitted our age to equal and surpass Marsh's age.

Friedrich Ratzel

The views held by a German geographer who had come to his studies through a doctorate in zoology and through early travels as a newspaper correspondent for a Cologne newspaper were quite different from those of the first American minister to Italy. Friedrich Ratzel (1844–1904) had submitted his thesis to Heidelberg on the anatomy and systematics of the Oligochaeta, an order of worms, and was able to pursue his zoological and natural history interests on his travels in Hungary, the Siebenbürgen (Romania), Italy, and Sicily. After the Franco-Prussian War, in which he was severely wounded, he came under the influence of Moritz Wagner, a famous student of nature, and Karl Alfred von Zittel, a paleontologist and geologist.[13] Travels as a correspondent for the *Kölnische Zeitung* increased his knowledge of Italy, Sicily, the Carpathians, and the Alps. He then undertook a trip to the New World through Cuba, the West Indies, and Mexico and across the United States to the Pacific, where he seriously considered settling down in California. His interests had turned almost completely to humans and geography and ethnology after 1875, and in 1880 he succeeded Oscar Peschell in the chair of geography at Leipzig.[14] In 1886 he received the chair vacated by Ferdinand von Richthofen, and it was in this atmo-

---

12. Glacken refers to an article that he describes as "a good summary of the work of the French chemist Jean-Baptiste Boussingault." The full reference is Richard P. Aulie, "Boussingault and the Nitrogen Cycle," *Proceedings of the American Philosophical Society* 114, no. 6 (December 18, 1970): 435–79.—Ed.

13. For more on Moritz Wagner, see the third essay in this book, "Darwin and His Contemporaries." Karl Alfred Ritter von Zittel (1839–1904) was a German geologist and paleontologist, professor of paleontology and geology at Munich, and director of the natural history museum there. He published several classic books on geology and paleontology.—Ed.

14. Oscar Ferdinand Peschel (1826–75) was a German geographer and anthropologist and professor of geography at the University of Leipzig, and well known for his work, *The Races of Man: And Their Geographical Distribution* (1876).—Ed.

sphere at Leipzig that Ratzel produced his famous works on political and human geography.[15]

Part of one of his most famous works, the *Anthropogeographie* (1891), was devoted to a delineation of the nature and extent of the habitable world. Ratzel was a voluminous writer, and his opinions are widely scattered throughout his works. But in the second volume of the *Anthropogeographie* he gives a detailed account of the habitable world, its development, its boundaries, its empty spaces, and its population.[16] This is of great interest as the expression of the conclusions of one of the great geographers of the nineteenth century. Friedrich Ratzel, said Joseph Partsch, himself a student of ancient ideas of the habitable world, was the first to grasp this idea and to give an exhaustive account of it.[17]

Ratzel himself carried the concept back to the ancients. The idea of the *oikoumene*, one of whose most common meanings was the habitable Earth, was of exceptional value—especially with the realization that the inhabited world of their time was but a small part of the planet and that the rest of the world was either uninhabited or uninhabitable. This useful distinction, Ratzel thought, had been lost in modern times, resulting in serious misinterpretations. The source of the error lay in modern views that the whole Earth was the dwelling place of man. This was an error. This is a central idea in Ratzel, as his ensuing argument makes clear.

One of the most important conditions for human evolution was being overevaluated if one assumed that the 510-odd square kilometers of the Earth's surface were available to mankind, when man could effectively occupy only two-thirds of this. Ratzel held the poetic phantasies of Johann Gottfried von Herder and Carl Ritter responsible for the view of the Earth as the home and schoolhouse of mankind.[18] He also took the leading cartographers of his day to task for making a similar error, stating that their error was not due to oversight but lay in a failure to appreciate the importance of the problem. Arguing further that "a perception of the concept of mankind is directly dependent upon the same perception of the idea of the oikoumene," he wrote that the habitable world is a fact that with growing knowledge could be observed more and more precisely. Historically, he said, the extension of the oikoumene since the age of discovery was one of the most significant of all facts, and "the history of discovery is the history of the oikoumene."

15. Ferdinand Freiherr von Richthofen (1833–1905) was a German geographer and expert on China who coined the term "Silk Road."—Ed.

16. Friedrich Ratzel, *Anthropogeographie* (Stuttgart: J. Englehorn, 1891).

17. Joseph Partsch (1851–1925) was a German geographer who succeeded Ratzel in the chair of geography at Leipzig.—Ed.

18. Johann Gottfried von Herder (1744–1803) was a German philosopher, theologian, and poet. Carl Ritter (1779–1859) occupied the first chair in geography at the University of Berlin and is considered one of the founders of the discipline of geography.—Ed.

The peak of these discoveries was reached in the scientific explorations of the nineteenth century, of which the beginnings in Africa were the most influential. With so much of the world now known, the implications of its insular character should be known as well. Ratzel wrote that the fact that basically all the land of the Earth is only an island in a sea four times as great naturally strengthens the concept of the isolation of the world in which one lives. Only in our times were the outermost boundaries of the oikoumene recognized and all mankind stood before us in its entire spatial distribution. This has produced the modern view of the nature of the world, one characterized by a sharper understanding of the relationships of ideas to the places and the areas of the Earth. Crucial to this understanding is the recognition that no people of any size occupy all the space over which it extends; these vacant spaces become smaller and smaller but never entirely vanish, and even the most populated parts of the world have room for more. Ratzel pointed out that eastern Belgium, Saxony, the Low Countries, Egypt, and Bengal, which constituted some of the most populated parts of the Earth, still had forests, moors, meadows, and coastal stretches to claim for cultivation.

He then turned his attention to environmental change, pointing out that deforestation had cut down on those vacant spots between peoples and that in the forested regions of both the Old and New World, cultivation was "practically synonymous with deforestation." He argued further that reducing the forests in the interests of expanding culture could easily lead to a destructive war whose goal was the complete destruction of the forests. Great transformations in the lands of historical peoples had come about this way. With the forests gone, there would be an economic loss, loss of protection, a worsening of climate and a deterioration of soils, an increase in floods, and landslides.[19] Forest conservation, he said, was essentially a modern phenomenon, and, more than any others, the lands of old cultivation had lost through deforestation what the intensive agriculture that took its place was unable to replace. Referring to North America, where he had spent quite a bit of time, he said that this deforestation was the reverse side of the great cultural progress. Forest destruction, he argued, was really an act of culture, of more dense population and better tools. However, though he stated facts and gave many historical illustrations from different parts of the world, there is not, as in Marsh, the sense that such practices were ruining the Earth. Perhaps this way of thinking was strange to Ratzel because of his faith in high culture and his belief in the Europeanization of the world. Besides, he had other fish to fry.

Ratzel's concept of the relationship between man and Earth as being primarily

---

19. Glacken discusses the topic of alpine torrents in detail in *Traces on the Rhodian Shore.* —Ed.

a spatial one in which human settlements, great or small, were scattered about on Earth and separated from one another by uninhabited areas naturally led to a study of population density and distribution throughout the world. The relationship of the distributions of the world's peoples to the physical Earth, Ratzel thought, rested on certain general observations: the boundaries of the inhabited world reaching to the uninhabited polar regions; the sparse populations in the trade winds belt of the Northern and Southern Hemispheres; the concentration of population in the continental regions in the Northern Hemisphere, especially in its temperate regions; the scattered occurrence of dense populations on the medium and smaller islands; the accumulation of populations on the oceanic rims and their paucity in the interiors; the dense population of interior lowlands, especially river valleys, in contrast to the more thinly populated uplands, the exceptions to this rule being in the mountains of tropical regions and the trade winds belt; and the growing dependence of populations of all countries on commercial centers and trade routes.

Crucially, however, he concluded that temperature, moisture, elevation, water, and climatic influences more generally conditioned the distribution of population, and claimed that great inequalities in the distribution of population could be observed even in the most overpopulated countries of the Earth, such as China and India. Ratzel then related dense population with high culture and argued that a decline in culture and decline in population went hand in hand. He offered the history of Europe as an example, stating that it had, in 1891, at the time he wrote the book, the highest population density of any part of the globe, and stood at the apex of culture. He theorized that periods of fast-growing population were closely related to general cultural development, and that this development in turn increased the carriers of the culture, which in turn increased the ability to advance and expand.

Looking next at population growth as a general human phenomenon, Ratzel claimed that the Earth is not nearly as densely populated as it might be. Mankind is still growing, still expanding, fertile lands are still available, while less fertile lands are supporting more than they are able to sustain. He claimed that the future would see much higher densities, and in some places a thinning out of population, until the goal of a distribution was reached in which in every part of the world the number of people was in harmony with their portion of the soil. This, he added significantly, would change not only the statistical picture of man's distribution but also the ethnographic, and with time, the potential for cultural enhancement.

Ratzel did not discuss the mechanism by which the world of the future would consist of populations adjusted to their environment, although he alluded to the historical emigrations of Europe as an example, arguing that the filling up of the Earth with peoples of European origin was "the most remarkable example of a population growth of great significance, a growth where the basic cause is the strong internal increase of European peoples in a limited area." In its power of breeding peoples, Ratzel argued, lay the most important reason for Europe's outstanding role in the history of mankind for the past two thousand years. For a great part of the history of the Earth, he contended, Europe had held, through its power of population increase, the position of "a superior, culturally strong mother land." The habitable world of the present showed these trends, and the future adjustments presumably would be a result of a more developed Europeanization. Europe had become to the whole world what Rome was to countries of the Mediterranean. The consequence of the dense population of Europe was the emigration to non-European lands, which would then be colonized, cultivated, and above all Europeanized. This emigration, he stated, was a pressing necessity for Europe, and the worldwide ensuing exchange would be based on sending food and meat to Europe in exchange for Europe's people and culture.

For Ratzel, then, the tendencies at work were toward an equalization of the population densities throughout the world under European leadership and with European immigration. The habitable world of the future would thus be a cultural creation of the European peoples. This Europeanization of the world was already far advanced: one could now set up a cultural scale of nations in which the most active were those where European influence had been been most prominent. At the top were the United States, followed by Canada, southern Australia, Africa, and South America. European influence was also apparent in North Asia, the Caucasus, Algeria, and West India, the islands of the Pacific, India, the Sunda Islands, Africa, and the Philippines. Moreover, Japan and Egypt had come under European influence without becoming politically dependent. Only a few countries, such as Morocco, Abyssinia, China, and Korea, had remained free of the influence. This was a convincing array, and any observer in 1891 could well look forward to the Europeanization of the world. However, time and circumstance have changed this likelihood, and, sixty-odd years later, it seems highly improbable that the habitable world will be dominated by one cultural tradition.

In an earlier part of *Anthropogeographie*, Ratzel had spoken of the various soils of the world and their importance; he recognized the black earths of South Russia, western Siberia, and North and South America as the reason for these regions' great wheat crops. Soil types showed a decided dependence on climate and had a zonal formation. In his discussion of soils, Ratzel stated that the most intractable of the questions to be addressed had to do with the capacity of soils for cultivation, which, as of his writing in 1895, had not been solved. He pointed out that under- and overevaluation stood sharply side by side in German East Africa, German Southwest Africa, and British Central Africa; whereas the steppe lands in western North America, which ten years earlier had been considered absolutely worthless, had attained value through the extension of irrigation. Again, the expression "Africa a new India" was formerly laughed at, whereas at the time of his writing the book, scholars claimed that tropical Africa, compared with India with respect to land alone, was richer—with its poverty explained by the ostensible fact that it had no industrious indigenous population. With the comment on Africa, Ratzel drove home one of his big claims, that what one calls the national endowment of a country is a many-sided affair.

At the base of the descriptions of the habitable world as a combination of nature, history, and culture lay two ideas of space, one as understood by a geographer, the other as understood by a biologist. There was a struggle going on in the world, a struggle for space, which should not be limited to the struggle for existence. Space is the final common limiting factor. This is set by the lands—not the Earth's surface, but of its available land. He wrote that anyone who wished to understand the history of life on Earth must at least appreciate this limitation, arguing that even in the greatest and most influential works of a biogeographic character among his contemporaries, what he called "this decisive basic fact, the limitation of space," was not even mentioned. The argument was simply this: every being, plant, animal, or human, needs space. These beings are organized in a unity, of which the successes of meteorology, climatology, and plant and animal geography give proof. Limited space intensifies the struggle for existence, and the area of the theater in which the struggle for existence is proceeding is of decisive importance. He argued that the struggle for existence is just as much influenced by the space available for it "as those high points in the armed conflict of men which we significantly call battles." In large areas opponents can withdraw; in small areas of struggle the struggle is desperate and decisive because there is no way out. The size of the place of the struggle is of decisive importance.

Ratzel claimed that Charles Darwin had not discovered the struggle for existence. Rather, he had shown in a profound manner its influence on the history of life; the importance of the power of organisms to increase geometrically, and, if not interfered with, any organism, including elephant or man, would gradually fill up the whole world. Unfortunately, according to Ratzel, Darwin did not pursue this question further, even though it was of highest importance. He argued that the organic world of our Earth is witness to the struggle on the entire life-supporting surface of the planet and bears the traces of this limitation. This, he claimed, applied to contemporary human society as well, so that "we live today in a time of lesser fullness of life and more restricted movement and closer relationships." When man's relation to the whole Earth is considered, he argued, his area for existence only lessens; it is not increased.

This view of the habitable world was a combination of two ideas: that the habitable part was much more limited than poetical fancy and cartography had allowed, and that the struggle for existence was transformed into a struggle for space by this elementary fact. Malthusian theory had been brought down to earth and made concrete. It is Darwin, of course, whom he mentioned, but it was the part of Darwinian theory that had been borrowed directly from Malthus that concerned him. The world was not expanding, it was shrinking; the areal gaps between peoples were being filled up, first by the struggle for space, second through the destruction of organic nature. This struggle for space had in the case of Europe produced a high culture that could be exported along with the emigrants to secure the final balance of population and space under a European hegemony. Perhaps here Ratzel failed to follow his own criticism of European-centered philosophies of history.

Ratzel's writings are difficult to piece together into a system as a whole. A sympathetic student of Ratzel's works has pointed out this difficulty and the tendency of Ratzel to speak in aphorisms. I think it fair to say that the view of the habitable world we have here is one of a relatively unchanging environment, in control of human affairs. The environment, especially a shrinking environment holds the last word, especially with the use of Malthusian doctrine, for which Darwin served as the intermediary. The habitable world remains the theater for human activity, just as it did for Herder and Richter, but it is a smaller, more sharply delimited, more carefully mapped habitable world. It is, in accordance with Ratzel's principles, a world that reaches for the soils and plant life through the political life of his time. And perhaps his own rejection of Bismarck, his adherence to the colonial policy of William II, and his advocacy of a greatly developed German navy were enmeshed in these more general ideas. His is not the habitable world of Marsh being torn away by human agency; it is a world filling up through the struggle for space, under the leadership of European culture, itself a product of that struggle.

E. G. Ravenstein

The worldview presented by Friedrich Ratzel was not concerned with an evaluation of the ability of the world to support its present or an increased population. He was more interested in ethnology, the distribution of peoples, the effect of the environment on civilization, and the struggle for space as it could be observed in those parts of the world suitable for human habitation. But toward the end of the nineteenth century and during the first few years of the twentieth century, up to World War I, there was a revival of interest in the maximum population that the world could support. It is difficult to assign any special reasons for this. Perhaps the revival of interest in Malthus had something to do with it, though he had attempted no computations of a maximum population. Godwin, in his answer to Malthus, had estimated that the world could hold nine billion people.[20] The great interest of European powers in colonial affairs, the widespread preoccupation with mapping, especially of Africa, and the feeling that nineteenth-century prosperity may have been temporary might have had something to do with the renewed interest in maximum population. Another reason might have been the shock, in England, of the 1865 publication of W. Stanley Jevons's *The Coal Question*, in which the Malthusian point of view was applied to the coal reserves, and the associated idea that British prosperity would collapse with their exhaustion.[21] Other reasons might have included the optimism and pessimism caused by the quick disappearance of guano and the uncertainties relating to the availability of the artificial fertilizers, a concern stimulated by Justus von Liebig.[22] Also, Sir William Crookes had delivered his famous presidential address before the British Association for the Advancement of Science on the dangers of the world's food supplies becoming exhausted, a warning based largely on the possibility of exhausting chemical fertilizers.[23]

In 1890, a German-born English geographer and cartographer, Ernst Georg (E. G.) Ravenstein, delivered an address at Leeds before a joint meeting of the Geographical and Economic Sections of the British Association for the Advancement of Science. Its title was "Lands of the Globe Still Available for European Settlement."[24] This was the first modern attempt to compare the present area and population of the world with the potential population that the Earth really could support. Exhaustive attempts had been made previously by others, in a series of volumes of *Petermann's Mittheilungen*, to compile from all sources an accurate estimate of the world's population and place it on maps.[25] Nevertheless, knowledge

20. See Glacken's discussion of Godwin in the first essay in this book.—Ed.
21. William Stanley Jevons, *The Coal Question* (London: Macmillan, 1865).
22. For an excellent contemporary book on this subject, see Gregory T. Cushman, *Guano and the Opening of the Pacific World: A Global Ecological History* (Cambridge: Cambridge University Press, 2013).—Ed.
23. Sir William Crookes (1832–1919) was a British meteorologist and spectroscopist known for, among other things, his discovery of thallium. His speech was a call to arms for industrial chemistry. See William Crookes, "Presidential Address to the British Association for the Advancement of Science 1898," *Chemical News*, 1898, 78, 125.—Ed.
24. E. G. Ravenstein, "Lands of the Globe Still Available for European Settlement," *Proceedings of the Royal Geographical Society and Monthly Record of Geography* 13, no. 1 (January 1891): 27–35. All subsequent quotations from Ravenstein in this section are from this work.
25. August Heinrich Petermann, *Geographic Releases*, 1st ed. (Gotha, Germany: Klett-Perthes, 1855). Augustus Heinrich Petermann (1822–1878) was a German cartographer and a professor at the University of Göttingen who revived the *Geographisches Jahrbuch* (Geographical Yearbook), edited by Heinrich Berghaus from 1850 to 1952, under the new title of *Mittheilungen aus Justus Perthes Geographischer Anstalt über wichtige neue Erforschungen auf dem Gesamtgebiet der Geographie von Dr. A. Petermann*. After his death in 1878, the journal was renamed *Dr. A. Petermann's*

of the world and exploration and discovery were two different things. Ratzel had spoken of the nineteenth-century contributions to the knowledge of the ecumene,[26] and although the history of discovery might be the history of the ecumene, this did not mean that the world's population could be accurately judged or that its lands available for cultivation could be evaluated with any precision. "The question which I have been invited to discuss with you today [September 8, 1890]," Ravenstein said in his opening remarks, "bristles with difficulties, owing to the paucity of facts which are at our disposal. In order to answer at all satisfactorily, I shall be compelled to determine: (1) The present population of the world, and its probable increase; (2) The area capable of being cultivated for the yield of food and other necessaries of life; (3) The total number of people whom these lands would be able to maintain. Speaking in the presence of so many able economists and geographers, I need not point out, that a precise answer to the apparently simple question is quite beyond my power."

Ravenstein approached this problem with a rich background in geography and cartography. Born in Frankfurt-am-Main, he had been a pupil of August Petermann and had worked in the topographical department of the British War Office from 1855 to 1872. In 1890 he was president of the Geographical Section of the British Association for the Advancement of Science, and in 1902 the Royal Geographical Society conferred its first Victoria Medal on him. In 1881–83 he published 1:1,000,000 maps of eastern equatorial Africa and in 1899 a 1:500,000 map of British East Africa. Aside from being regarded as an authority on Central and East Africa, he had studied deeply the historical geography of the Middle Ages, the age of discovery, and the history of cartography. Ravenstein could thus speak as a master of contemporary and historical geography.

The difficulties began at the outset: estimating the world's present population was basic to the inquiry,

> but it is quite impossible to reply to it with any amount of confidence. Enumerations of the people have been made in all civilized states, but with respect to large parts of the world we are still completely in the dark. Of Africa we know next to nothing. Whilst the long array of figures presented to us as the result of a census taken in China are not calculated to inspire confidence. I have taken some care to form a true estimate

Mitteilungen aus Justus Perthes' Geographischer Anstalt and later, in 1938, *Petermanns Geographische Mitteilungen.*—Ed.

26. Ecumene is a term used by geographers to refer to the inhabited portion of the earth, including those areas used for work and agriculture, as well as domestic residence. The terms "oikoumene" and "ecumene" are used interchangeably in this work, depending on the original author's preference—Ed.

of the population of Africa, and I cannot believe that that estimate supports more than 127 millions, instead of the two, three, or even four hundred millions allotted to it by certain statisticians.

Ravenstein in his address then attempted to break down the land area of the world—excluding the polar lands—into three types of regions: fertile, steppe, and desert. This method yielded an overestimate of the fertile regions since "it cannot be assumed for an instant, that the whole or even the greater part of it could even be converted into fields yielding the fruits of the earth. There are within it mountains, which will never tempt the agriculturist, sandy tracts; capable of supporting only forests and even steppes or poor savannahs, not fit for anything except the raising of cattle." There was a similar lack of uniformity in the steppe regions (or poorer grasslands), "and as within the 'fertile regions' we meet with comparatively sterile tracts, so within these 'steppes' there exist large areas which can be rendered highly productive, especially where means for irrigating the land are available." The third region included the deserts, "within which fertile oases are few and far between." The acknowledged grossness of this estimate is apparent because of the unknown quantity of the minority lands in each category of fertile and steppe. Nevertheless, out of the total 46,350,000 square miles—excluding the polar regions—of the Earth's surface, 28,269,000 are classified in the fertile category. This fertile region represented slightly over 61 percent of the total area, compared with Ratzel's estimate of two-thirds of the world's area for human use. The full significance of this can be seen when we combine the estimate of the area of fertile regions with the population as of 1890, as presented by Ravenstein:

	Area of fertile regions (square miles)	Population in 1890
World	28,269,000	1,467,600,000
Europe	2,888,000	380,200,000
Asia	9,280,000	850,000,000
Africa	5,760,000	127,000,000
Australasia	1,167,000	4,730,000
North America	4,946,000	89,250,000
South America	4,228,000	34,420,000

The great areas of hope were Africa and South America. Ravenstein had now completed two of his estimates. Now, how many people would the Earth support "once it had been finally brought under cultivation"? Here he ran into the same difficulty that the better equipped investigations of today face. Would this be based on a vegetable or an animal diet?

> I observe for instance that there are present some vegetarians. These, if their opinions were asked, would maintain that if man returned to nature, and fell in with their peculiar views, three men could live where one lives now, and six men might take the place of one of our larger domestic animals, which would, or course, become extinct once their dietary value became a thing of the past.... I am not sufficiently utopian to believe that mankind generally will ever be prepared to accept these principles, or that, having accepted them, man would not degenerate.

Could not the present areas of the world through more efficient methods be made to support an increased population without increasing the cultivated areas? "This no doubt is true as respects to many countries," he answered, "but it is hardly true of the world at large." The agriculture of the newly settled areas, including the United States, had been very wasteful and pointed toward immediate returns.

> If you travel from Montreal to Washington you pass through millions of acres of land, which were once most productive and are still lovely to look upon, but which nevertheless produce nothing. The forests have been devastated in the most reckless style, and swamps and sandy waste have taken the place of trees. These things, however, will be mended in the course of time; the exhausted soil of the eastern states will recover; and the forests, where wantonly destroyed will be replanted. In proportion as the population increases, so will the resources of the country be more carefully husbanded. Of course, when preparing my estimate of the possible populations of the globe, I assumed that the available areas would be rationally cultivated, and I even admitted a slight improvement in the yield of each acre.

There remained the crucial question of the standard of living and its relation to the capacity of the globe. Ravenstein found, of course, that this varied throughout the world, and that it was necessary to take representative peoples. Continental Europe, for example, from the North Sea and the Atlantic to the Baltic supported 156 inhabitants per square mile. Countries like Germany and France were importers; Hungary, Romania, and Belgium had a surplus. Food deficiencies in this area could probably be made up by these countries of southeastern Europe, so that the average of 156 per square mile could stand. In Asia, India supported a population of 175 per square mile, China 295, and Japan 264. Ravenstein remarked: "Of course, the standard of life in these countries is different from what it is with us, and it is not likely to undergo any material change in the near future, for diet is as much a question of climate as it is of inherent disposition." If these countries are taken as representative of the fertile regions, their mean population is 207 to the square mile. "If I accept this figure as approximately the truth," he argued, "the 'fertile' regions would be able to support 5851 million human beings, and this I believe to be a moderate estimate." Without further elaboration, he estimated that the steppes could support ten inhabitants, "whilst the deserts would be fully peopled if they had even one inhabitant to a square mile. The total possible population would consequently amount to 5,994,000." It will be of further interest to add here what Ravenstein did not supply, a picture of the population of a filled-up world.

For the Europeans, Ravenstein held out little hope for tropical settlement. To render tropical countries fit places of residence for European colonists it would be necessary either to change the constitution of Europeans or to bring about a change in climate. A third alternative, for Europeans to change their culture for tropical climates, Ravenstein did not mention, but perhaps this is more efficient than either of the other two! But the fact that the tropics offer no opportunity for European emigration does not mean that there should be corresponding departures from the Earth's total carrying capacity, for the reason that "it is not necessary that the consumer of food should live in the country which produces it." This will depend on world trade, and the food supplies of the world will increase in direct proportion to the development of this interchange. He said: "A time will surely come when the millions densely crowded together in the temperate regions of the world will draw a large proportion of their food supplies from tropical countries, which at the present time just manage to maintain their own scanty populations, and sometimes not even that." In other words, the tropics are the hope of the world. The peoples of the Earth need not redistribute themselves; the fertile regimes will supply them in absentia, like absentee owners. But there is no detail here about the shipping lanes, the road networks, the railroads, and most of all, a formidable international situation to feed the world's peoples. Seen in this light, a world carrying capacity of about six billion is subject to more ifs, ands, and buts than Ravenstein had concluded. This world population Ravenstein calculated might be attained 182 years hence, or in 2082.

In the discussion that followed the presentation, the economist Alfred Marshall, perhaps thinking of the research Jevons's work had started, said that Ravenstein had not mentioned fuel.[27] Many felt that checks on population growth in the future in the temperate zones would come from a scarcity of fuel, not food, while others denied the danger. This would pass the problem on to the physicists, and even if one had all the facts, a paraphrase of Marshall's remarks continues, "supposing they were careful not to overpopulate, how were they sure that the world would not be overpopulated by people who were less careful, and whom for that very reason,

---

27. For an account of Marshall's views on population, and his response to Ravenstein, see J. J. Spengler. "Marshall on the Population Question: Part II," *Population Studies* 9, no. 1 (July 1955): 56–66.—Ed.

perhaps, the world would less care to have?" Other comments ranged from an optimistic view that as the ultimate limit was approached, greater attention would be paid to all forms of economy, and population growth would proceed more slowly, to concerns about population explosion and control, to ideas that emphasized the need for improved cultivation and improved transit.

I have analyzed Ravenstein's short, almost forgotten talk in some detail because it asked, and attempted to answer, a type of question that is of major concern today. When Ravenstein posed it, it was a new approach, especially the assumption that a more rational use of land and national resources would, as a matter of necessity, accompany a population increase. Unlike Marsh, however, Ravenstein believed that with time, the destructiveness of the past could be corrected. Like Godwin before him, and like all present students of the subject, he had to grapple with the standard of living of the various peoples of the world in relation to the Earth's capacity, which he assumed to be a fairly persistent and unchanging factor through time. His estimate also assumed there would be no deterioration of the world's resources. But most fundamental of all was the idea that this could be achieved only through worldwide exchange and a highly developed transport system. The problem of appraisal in terms of diet also remains with us to this day. Much of the present-day food resource appraisal must be done in terms of caloric requirements either from a vegetable or a mixed diet. It is chastening also to recall that toward the end of the nineteenth century—as now—with all the geographic exploration and statistical inquiry, the materials for assessing the resources of the world were so meager. And if one should smile at the grossness of this estimate—in which all unusable land included in the fertile regions was multiplied by 207, and all fertile lands in the steppes multiplied by 10—it would serve us well to recall that recent attempts at similar estimates have required equally gross multipliers. Salter, for example, in 1946 estimated that 1.3 billion acres of the world's soils could be brought under cultivation: one billion acres in the tropical and subtropical areas of the world and 300 million acres of the world's soils.[28]

Carl Ballod

The work of Ravenstein stimulated considerable interest in the question of the Earth's carrying capacity. The Prussian statistician Arthur Freiherr von Fircks thought that Ravenstein's multipliers were too low, and, basing his calculations on a vegetarian diet, concluded that, if it had to, the Earth could hold 9,272 million, assuming that all fertile land was cultivated intensely and exclusively devoted to food

---

28. R. M. Salter, "World Soil and Fertilizer Resources in Relation to Food Needs," *Science* 105, no. 2734 (1947): 533–38.

production, that the steppe regions were optimally used for cattle breeding, and that the sea and inland waters were used for fishing.[29] Fircks's conclusion seems to summarize the attitude of European scholars of the time, namely, that the danger of global overpopulation was far away, but real for some European states.

The attitude of an overpopulated Europe in an underpopulated world is clear in the attempt made by the German political scientist Carl Ballod (Kārlis Balodis, in his native Latvian) in 1912 to estimate the number of people the Earth could feed. His work has interesting departures from that of Ravenstein, even though the original inspiration came from him.[30] The problem of estimating on a world basis, Ballod said, was closely tied up with the population problem, population policy, and one's attitude toward Malthusianism. The world's peoples could be roughly divided into those, such as the Chinese and the Japanese, who existed on plants, with a minimum dependence on animals, and the peoples of the New World (the United States, Australia, Argentina, Uruguay), whose lands were largely given over to the feeding of animals. If the Japanese, the Germans, and the Americans were taken as examples of the need for grains (based on a vegetarian or an animal diet), the Japanese would have about 200 kilograms a year and the Germans 450, while the Americans would need 1,000 kilograms per year per person, based on a higher use of meat, milk, and butter. There were some caveats. Lands that had to be devoted to the clothing of the world's peoples needed to be considered. Crucially, however, Ballod, like others before him, was concerned with the problem of exhaustion of fuels in relation to the habitability of the temperate zones. It was fortunate that there was coal in this period of history. If this were not so, significant portions of the Earth's lands would have had to be given over to wood production. Moreover, one could proceed from the assumption that in the far distant future, when its coal had been exhausted, humanity would prefer living in the subtropical lands, where little or no fuel is required for winter warmth, and consider the regions of the temperate zone merely as areas for the production of agricultural products. Another concern was with artificial fertilizers. The Japanese and Chinese had kept up production for at least two thousand years with intensive methods and no artificial fertilizers. Europe's high production had been the result of favorable soil and climatic factors and artificial fertilizers, but it was by no means certain that the supply of phosphoric acid needed for plants would last forever. This fear of phosphorus exhaustion appears frequently at the turn of the century. Under these cir-

---

29. Arthur Freiherr von Fircks, *Bevölkerungslehre und Bevölkerungspolitik* (Leipzig: C. L. Hirschfeld, 1898), 295.

30. Carl Ballod, "Wieviele Menschen kann die Erde ernahren?," *Schmollers Jahrbuch für Gesetzgebung, Verwaltung und Volkswirtschaft* 36, no. 2 (1912): 81–102. [Kārlis Balodis (1864–1931)/Carl Ballod was a Latvian economist, statistician, and demographer.—Ed.]

cumstances, Ballod argued, it was of greatest importance to first estimate how the soils of the Earth were divided among the various nations, and then to determine how much land could be brought under cultivation. However, he argued, the extant knowledge did not allow for such a determination to be made with any accuracy. This would have required an exact land classification and soil evaluation of the lands of the world, as well as the determination of climatic relationships.

Here, Ballod proposed an interesting methodological innovation. He argued that knowledge of climates was sufficiently good to enable the determination, with some accuracy, of those types of lands suitable for agriculture or forestry. A close examination of climatology, he proposed, could therefore offer a provisional substitute for the overall evaluation of soils, which was lacking.[31] In making this proposal, Ballod accepted a range of error of 10 to 20 percent, which, he claimed, was a bargain. Ballod thus made climate a surrogate, a measuring rod for a land's ability to produce, for unlike soils, which can be improved, humanity can do nothing to change the climate except in a local or a minor way. Pursuing this theme, and after considering temperature and rainfall in various parts of the earth, Ballod concluded that the tropics, regions with over 2,000 millimeters of annual rainfall (about 79 inches), were "double value," since they could produce two harvests a year on the same soil. As Ballod saw it in 1912, the tropics, with about 600 million people, had 30.6 million square kilometers of land suitable for cultivation. In the temperate zones, with a corresponding 1,130 million people, there were 25.3 million square kilometers. The great areas of expansion therefore were in the African and South American tropics. The latter, according to him, had, of all the parts of the world, besides Europe, the most favorable "coefficients" for arable soil. The Amazon basin almost alone could feed the entire population of the Earth. Ballod's discussion of the tropics shows that the European races at that time were in a dilemma regarding the tropics. They were unsuitable for them, yet it was these regions in which lands one and one-half or two times the fertility of the temperate zones were to be found. And in the future, the European races might treat the tropics as areas of exploration, or might come closer to adapting themselves to the tropical climate. Such lands could also serve in the future as agricultural regions for the lands of the temperate zone.

Ballod made three estimates of the maximum number of people the Earth could hold. He estimated the total available land at 56 million square kilometers or 5,600 million hectares, of which he assumed 2,800 million hectares would be used for food production. The American standard of living required 1.2 hectares (almost 2.471 acres), including about 0.9 hectares per person for land planted to grains, and

31. Ibid., 88.

the rest for food and draft animals, cotton for clothing, and so on. At the American standard, the total possible population of the Earth would be 2.33 billion (2,800 million hectares ÷ 1.2), which, at the rate of increase at the time Ballod was writing, could, he projected, easily be reached in a generation. Again, considering only the German standard of living, which required half a hectare per person, Ballod estimated the total capacity to be 5.6 billion. And at the Japanese standard of living, even allowing for the lighter weight of the people, the requirement for food owing to intensive agriculture would be 0.125 hectares, and at this standard, 22.4 billion people could live on Earth. Ballod thus estimated that the Earth's population could go no higher than 5.6 billion if any sort of standard of living were to be maintained. An increase over that number would be conceivable if a lasting supply of phosphoric acid were available; the number might then double.

As a German professor of political science in 1912, Carl Ballod was able to draw some conclusion pertinent to his own country. Taking everything into account, he argued that it was possible to triple the world's population while retaining a high standard of living. Translating this into national strategy, he stated that the German position should be guided by the idea that only those people who possessed a strong ability to increase had a future and that only they would rule and divide up the world among themselves. Malthusianism thus was not the only population doctrine that could be used as an excuse for conquest. Germany, compared with England, France, and Russia, had, he argued, historically suffered a stepmother's treatment. While insisting that force was not the solution, he argued for the principle of the distribution of land according to actual population.

Aside from relating his discussion to German national policy, Ballod's estimates are of interest for their frank recognition of the relationship of living standard to the Earth's power to feed its peoples, and the danger to those with a higher standard of living if the discrepancy between the highest and the lowest became too great. Before becoming a political scientist Ballod had been a geographer; while he gives no details of his calculations, his estimate is obviously based on climate and a classification of standards of living. The question of the exhaustion of phosphates obviously bothers him most, and it is not possible to tell from his paper whether he made use of the technical soil literature. But most striking of all is his focus on the tropics as the real keystone in his analysis. In that respect the situation has not changed to this day, for one of the chief areas of difference between optimists and pessimists as to the capability of the Earth to support its growing populations lies in in the divergent opinions regarding the development of the tropical lands of the world.

More than any of us realize, the ideas held about the tropics, especially the South American and African tropics, have had a great deal of influence not only on geographic thought but on the more general stream of modern thought. For a long time, the attitudes of the first settlers in the tropics of the New World colored thinking about these areas as regions of death and disease. In my own view, the writings of Humboldt did much to dispel this view and teach a love for the scenery of the tropics, a respect for its luxuriance—its fertility being assumed from this—and its baneful effects on ambition because of the ability to get a living so easily. It is worth mentioning here that Malthus had argued that the fertility of a soil did not necessarily mean the creation of wealth. In the late nineteenth and early twentieth centuries, a prominently held view, as we have seen in Ravenstein and Ballod, was that the tropics were permanently unsuited to white settlement. Ellsworth Huntington's writings also have been very influential in portraying the tropics as places of low energy, where high civilizations could not be expected, though in his later writings he relented slightly on this point. Gradually, however, with the progress of tropical medicine, the view has changed, and the problem is now regarded as one of acculturation rather than of a permanent physiological difficulty inherent in the peoples of Europe or of the temperate zones. Tropical luxuriance also, as we have seen, has been identified with tropical fertility, and more recent debates have centered on the fertility of tropical soils. Ratzel said as far back as 1899 that the fertility of tropical soils was an illusion. Many tropical soil scientists today point out the same thing. But wherever the truth may lie, the works we have reviewed were not based on knowledge of tropical soils, for even to this day such knowledge is scanty and spotty.

Edward M. East

In these earlier attempts to estimate the absolute number of people the Earth could hold, there was a large element of unreality. The totals thus arrived at had to be distributed somewhere, and they were. The regions, the areas, the continents, like good housewives, were meant, in the following decades or centuries, to obediently play the part assigned to them under European leadership. But looking at them now, one is immediately struck by the immense social and political changes that their consummation implies. Humanity seems like a fast flood that will flow over the Earth to fill in the empty spaces. Of course, this was not precisely true, for the argument was, as we have seen, put in another way too. The Earth could support so many billions of people; it was not necessary that they be supported where their food was grown. The existing great inequalities of population distribution could be maintained. The baking oven did not have to be in the middle of the wheat field. But this conception implied a permanent division of the world into agricultural and industrial areas, an unrestricted international trade network, and a well-functioning internal transport system.

Two wars have made these preconditions much more remote than they were in Ravenstein's and Ballod's time. After World War I, the bases for discussion were on an entirely different level. For one thing, interest in Malthusian theory revived, and Keynes's *Economic Consequences of the Peace* had in dramatic and forceful language called attention to the "Malthusian devil."[32] War and the threat of war always seem to produce some kind of a population theory, generally a refurbished one, for there is not room for any fundamental novelty. All during the twenties, problems of population were discussed. Some of these were concerned with world trends as a whole, whether the growth was projected as following Malthusian lines or as reaching an equilibrium; some were concerned with unequal growth by race, a point of view that produced the "yellow peril" literature.[33] And many of these discussions were also simply discussions of population by itself. Few discussed population with a proportionately thorough study of the Earth's resource capacity.

Edward M. East was among those who did. *Mankind at the Crossroads*, which he published in 1923, was a widely discussed book at the time, and parts of it, pieced together, give us a different view of the habitable world than any which preceded him.[34] East had been led to study the question of the present and future habitability of the world by his earlier interest in food chemistry, dietetics, soil chemistry, crop breeding, crop production, agricultural economics, plant breeding, genetics, heredity, evolution, and "the problems of the race in the collective sense."[35] At the time the book was published, he was professor of experimental plant morphology at Harvard University, later, in 1926, becoming professor of genetics. His study of human reproduction, he said, was first carried on without reference to the ability of a population to feed itself. But wherever he turned, the broader aspects continually asserted themselves. They could not be evaded. And from their consideration a startling truth emerged. The facts of population growth and the facts of agricultural

---

32. John Maynard Keynes, *Economic Consequences of the Peace* (New York: Harcourt Brace, 1919). [For a useful analysis of Keynes's theories of population, see William Petersen, "John Maynard Keynes's Theories of Population and the Concept of 'Optimum,'" *Population Studies* 8, no. 3 (March 1955): 228–46. In a note here, Glacken alludes to a chapter on Keynes that is now lost.—Ed.]

33. The term "yellow peril" was coined by Kaiser Wilhelm II of Germany in 1895, and held that people from East Asia threatened Europeans. A vast literature, ranging from fiction to opinion, emerged in the late nineteenth century onward and helped spawn racist xenophobia all over the Western world for at least half a century.—Ed.

34. Edward Murray East, *Mankind at the Crossroads* (New York: Scribner, 1924). Subsequent quotations in the text are from this edition.

35. Glacken notes here that "with the 'quality' part of the population I am not concerned." Readers may readily see his viewpoint by reading East's chapter on *Racial Prospects and Racial Dangers*.—Ed.

economics pointed severally to the definite conclusion that the world was confronting the fulfillment of the Malthusian prediction here and now. There was a choice to be made: whether man would control his destiny or be "tossed about until the end of time by the blind forces of the environment in which he finds himself."

The tardy realization of the predicament, East felt, had resulted from a divorce between the biological sciences and the social sciences, especially history and biology. Like Malthus, he thought the world's history had been written by the wrong people, referring to Malthus's claim that the principal reason why this topic had been relatively ignored in history was that histories of humanity thus far had only been about the higher classes. The idea of the evil consequences of a continued population growth, East held, was an old one, and demonstrable statistically.

East went on to argue that it was an error to consider population alone, and that full use should be made of facts of geography and present-day agriculture. He advocated for "a careful survey of the prospects" since the world was ostensibly fast reaching the end of its land reserve, and the end was in sight. In making this appraisal, however, East emphasized that there was no point in becoming frightened at the prospect of the white race being overwhelmed by other races. It should not therefore mistakenly undertake a population race with other peoples of the world for survival. The white race, he argued, "will survive, simply because the other races have no room to expand."

East argued further that the effect of mechanization of industry had been completely misunderstood, and "had not the slightest effect on the food requirements of the human animal." Mechanical invention, he said, probably had not increased "agricultural production by a single grain of wheat. It may even have decreased it." The reason was that most food was produced by intensive hand labor with a few simple tools. The increase in food production in the nineteenth century, he argued, had owed to the expansion of lands, not to the intensification of agriculture. Steam shovels, power plows, improved reapers and loaders, new methods, and cheap transportation had permitted a worldwide increase in food. "It cannot be too strongly urged," he wrote, "that the age of steel affected the population question solely because the world as a whole is still undermanned because there still existed large tracts of unbroken wilderness." The contribution of science lay in its power to make use of new land for an increasing population. "The prerequisite for further expansion, therefore, is new land; and new land is limited, decidedly limited."

The point East makes here represents an important effort to bring Malthusian theory, in an essentially unaltered form, into harmony with the progress of science and mechanical invention. It will be remembered that many writers of the late nineteenth and early twentieth centuries had argued that it was this progress in invention and science that decisively refuted the theory. To East, the contrary was the case. Population still tended to outstrip food supply. All science was doing was increasing the food supply, mostly by the use of new lands, and to a lesser extent by improving those already under cultivation, and in doing so permitting a still greater increase in population. As far as the ratios were concerned, therefore, science and mechanical invention could be regarded much as one considered the food supply: no matter what the improvement, there would be always be more than enough people to share in it.

Neither had the great international migrations of the nineteenth century affected the saturation in any fundamental manner. New colonies had been set up, and had increased and multiplied, largely because of cheap transport. However, he claimed, "in every commercial country Nature's banquet-table was . . . enlarged," in turn causing "an increase in the birth-rate," so that "the newly created seats were soon filled with hungry occupants." This raised the following question. With this rapidly expanding world's population, what did the Earth have to offer, and how long would it be before the saturation point was reached? East began by noting that the land area of the world, excluding the arctic regions, was 33,000 million acres. To determine the maximum limit of arable land, he used a statistic developed at the time by the International Agricultural Institute at Rome: in the most populous countries (and these generally had the highest percentage of arable land), the cultivated area averaged above 40 percent of the total area. If this figure was applied to the whole world, "13,000 million acres are available for food production."

East then estimated that each person on the globe would require 2.5 acres. This estimate, based on "a rather extended study of this matter" (the details of which are not given), rested on several assumptions: the existence of "sane beneficent governments," "adequate means of distribution, constant efficient effort equal to that of Western Europe during periods of peace, agricultural production equivalent to a return per acre midway between the average and the best in the world today, and a standard of living on a parity with what is found in the more densely populated countries of Europe." With these assumptions, it is clear that affairs would have to go very smoothly to support the world's peoples on an average of one per 2.5 acres. With these two figures, the maximum capacity of the Earth becomes a problem in simple division: 1,300 million acres ÷ 2.5 acres per person = 5,200 million people. This total is remarkably close to Ballod's estimate of 5,600 million, based on the German standard of living, which is perhaps not too far from the standard of living advocated by East.

Here, however, the similarity ends, as is apparent when Ballod's figures are converted into areas.

	Acre requirement per person	Food acreage	World population capacity (millions)
Ballod	1.2355	6,918.8	5,200
East	2.5	13,000.0	5,600

This is an instructive comparison, and transcends the estimates of the two men involved. Even granting the validity of calculating an average acre per person figure for the whole Earth, it is quite apparent that the total is a mere quotient, derived from two highly shifting basic figures. This is so childish that it would not be mentioned, were it not for the fact that it is the quotients or the total that are most often cited and accepted or rejected. In this case, just about ten years before East, Ballod drew optimistic conclusions from data that, if anything, were more pessimistic. From his figures, East concluded that the figure of 5,200 million would at present rates be reached in just a little over a century. He wrote: "Let us emphasize this result. Under the most optimistic assumptions as to production and distribution of food that it is reasonable to make, the world can support but 5,200 millions of people; and these people must content themselves with the limited dietary and the few material necessities which form the current standards among the peasantry of Europe." The peasantry of Europe! Ballod had given them a German standard of living with over six billion acres less to live on. It is not a matter of disparaging these men or of tiresomely analyzing out-of-date estimates. This type of problem is with us today, and it is the conclusions drawn from such estimates that furnish the optimistic or the pessimistic appraisals. So the world has 50, 30, 20, 10, or 2 billion of acres of land with 10, 5, 2, or one acre for each person. But where are they? Are they in Africa, the Amazon, in northern Canada, the Sargasso Sea, or in the Kalahari Desert?

East regarded his estimate as an ultraconservative one, that is, that it was the most optimistic statement of the case that could be made. He assumed that a future population would derive about the same proportion of its food from the sea as the world's population of his time did. He considered plankton as a source of food, but concluded "that proper data do not exist by which to estimate accurately the stock in old Father Neptune's larder." He also included a map of the world's arable land, but it was so generalized as to add little to the weight of his argument. East then turned to an examination of the areas of the Earth's surface that still could be settled and cultivated. These were in the main confined to Africa and "the Americas below the Rio Grande." But what were the obstacles here? Essentially, they were the ones that Ellsworth Huntington had been pointing out for a generation. It is interesting to see in the following quotation a curious reversal in present-day thinking regarding food and acclimatization in the tropics: "Food is produced rather easily by Mother Nature in the tropics, it is true, but on the whole the hot countries are not fitted for huge populations. Great heat evaporates energy as fast as it evaporates water, and this factor alone warrants us on not being too sanguine about filling up the tropics rapidly." Moreover, disease, the storage and transportation of food, epidemics, and plagues cut down the safety margin for the great tropical populations. East believed science would overcome these difficulties; but his real point was that the white races would never develop the tropics as places of settlement till stark necessity forced them to do so, writing: "When the time comes, he will go, but he will not set out to till grim necessity drives him on." The high tablelands of tropical Africa, on the other hand, were the most suitable for colonization in the interior period until the expansion of population to the limit forced the development of the tropics. This, he argued, "gives a respite of less than half a century at present rates of increase, of less than a century at the rate of increase which probably will ensue during that time." He concluded that "throughout most of the period the severity of the struggle in these countries will be marked."

In East's view, the future nature of the habitable world was almost entirely in the hands of the white race. It "is increasingly rapidly. Why? Simply because it has political control of nine-tenths of the habitable globe, and because it has the ability to utilize the space it holds" (116). He argued that any check would come from within the whole race itself. He wrote:

> The white race, therefore, will be limited in the immediate future only by its efforts toward a superior culture that will hold away the results of overpopulation.... At the end of this century, the whole race will probably number between a billion and a quarter and a billion and half. It will fill the temperate regions and the better parts of the subtropics. And the combined colored races of the world will not equal theirs in numbers. And there will be no international migration.

East chose to study the habitable world from the point of view of races, and the effect of this and other passages in the book is to imply a unity of the white races. But, as should have been clear from World War I but is abundantly clear from World War II and its aftermath, the habitable world will not be contested or fashioned by racial blocs acting in unison but by combinations of national groups, in which racial differences are the rule rather than the exception.

Part of East's thinking was based on theories of climatic determinism, especially as climate was deemed to affect the comparative abilities of races. The "climatic limitation of initiative," which was to be observed especially in the tropics, East regarded as "a remarkable thing. It may account for more of the white man's success in life than one would like to admit." Regions of progressive ideas have been regions of stimulating climate, and only the northern Chinese and Japanese today, among the black and brown races, live in areas of high initiative.[36] This view of present and future notions of the habitable world is thus a combination of (1) the original Malthusian theory, (2) a reconciliation of the theory with the success of science, and (3) a form of climatic determinism that ensures the domination of the peoples of the temperate regions, largely members of the white race.

In the world of the future, molded by these ideas, he believed, certain definite consequences would emerge. Population pressure would cut down on animal food. "Food animals, fowls excepted, are luxuries which must tend to disappear with an increasing population." Moreover, he noted that nothing could be expected from new plants, and that "every single acre of the important plant-food was discovered and brought into cultivation by prehistoric man.... If anything the needs of a saturated world will require fewer plants." He went on to predict that increasing population pressure would bring about the use of fewer crops. Those who saw in scientific plant and animal breeding new hopes for the future were comforted by the concession that the net gain "will permit the human race to pack together a little more tightly on the face of the earth, if this be any advantage. However, he cautioned, "the dark cloud of stored troubles will loom as forbiddingly as ever."

Soils and climates, East argued, were "the true arbiters of food production. Little could be done about climate. Soil, therefore, was the limiting factor determining the size of the world's population, the extent of its conflict, and the height of civilization. Soils differed in fertility, and "particular methods of treatment, may enhance or lessen their productiveness," he argued, and concluded that on the proper interpretation of these acknowledged facts depended "the whole future of the human race."

East's concept of the soils was that of an agricultural chemist, especially the elaborations that grew out of the theories of Liebig. Liebig's theory was, he claimed, "an exemplification of the old adage, you cannot eat your cake and have it. If you use up the natural fertility of the soil, it is gone. Four minus two is two." It was soil exhaustion, without fertility replacement with manures, night soil, or artificial fertilizers, that was reducing the food-producing potential. Despite centuries of cultivation, only recently had Europe strained its soils. Europe as a whole, he pointed out, "has almost *tripled* in population during the last century. Europe, then, is only now being cropped to death; her soils have *not* really been tested out for centuries." The need for replenishing soils was therefore urgent, and here the chief worry was about phosphates. In speaking of these problems, East used the American experience, but was its soil exhaustion typical of world conditions? (195).

---

36. East cited Ellsworth Huntington favorably here, stating that Huntington had disposed of the threat of black-brown world domination in his work, "Civilization and Climate" (Ellsworth Huntington, *Civilization and Climate* [New Haven, CT: Yale University Press, 1915]).—Ed.

To release the pressure of multiplying populations on soils whose fertility was exhausted or maintained only by constant enrichment, East urged worldwide birth control rather than trying to raise the food supply, which in turn would only increase the population of the Earth. "The vessel," he said, meaning the Earth, "is somewhat elastic, it is true, but it is a closed vessel nevertheless. Enhanced efficiency and prudent conservation of resources are means of bringing about increased happiness only if numbers remain constant." Science might be a stimulant to the creation of new resources, he argued, but it offers no real solution. Mechanical invention, by vastly enlarging the food return per unit of manpower without noticeably raising the return per unit of area, he said, has merely served to shorten the time when the Earth as a whole is saturated with people: "It is a vicious cycle. We raise more wheat to feed more men to raise more wheat to feed more men." Worldwide birth control held out the solution not only to the resource problem but to the eugenics problem as well. The birth rate must come down "to fit a rapidly diminishing food reserve," and "it must come down throughout the whole population and not merely within one section which furnishes those of greatest social worth."

The habitable world of the present and the future that emerges from Professor East's analysis is an ingenious combination of several ideas whose history has already been discussed. The basis lies in the Malthusian doctrine as a fundamental sociological theory. It was not that the ratios were accepted, for East accepted Pearl's theory of the growth of population describing a sigmoid curve.[37] This was not quite consistent with his Malthusian position, since, following Pearl's formula; the population would reach an equilibrium. But East thought of this more as postponing the point of world saturation "since the growth-curve will be characterized by ever-diminishing increments during the latter part of its course." Nevertheless, these considerations did not negate the main principle, that the population was growing far ahead of the food supply. The traditional assertion from Saint-Pierre claims that science could find a way out of the difficulty. East responded by subordinating science to the Malthusian doctrine, and claiming that mechanical invention was an agent of the postponement rather than of the solution to the world difficulty. To this stern doctrine East added an equally intransigent theory of climatic control that had been elaborated so successfully by Huntington. This emphasized the cultural inferiorities to be expected in tropical regions, the undesirability of the tropics for habitation except *in extremes.*

In a nutshell, East argued that the main struggle would go on in the already

---

37. Raymond Pearl (1879–1940) was an American biologist. See H. S. Jennings, *Biographical Memoir of Raymond Pearl 1879–1940, Presented to the Academy at the Autumn Meeting, 1942,* National Academy Biographical Memoirs 22 (Washington, DC: National Academy of Sciences of the United States of America, 1942), 295–347.—Ed.

populated areas of the world until saturation was reached, after a respite in Africa. To this he added a conception of soils that differed little from Liebig's. The emphasis was on exhaustion and loss of fertility consequent on increases in population. When to the Malthusian idea was joined the idea of the inhabitability of the tropics and the idea that soil exhaustion was creeping up on the world, the result added up to a belief that every measure was futile unless worldwide birth control was instituted. What is most surprising in this work of a biologist is the absence of an ecological point of view and the confinement of soil problems to those of exhaustion. The problem of erosion was not mentioned, nor was there any hint of the cumulative effects of occupancy that Marsh had dwelled on. It was a view whose origins are to be sought in Malthus, Galton, Liebig, and Huntington. These were the four horsemen of Professor East's apocalypse.

Albrecht Penck and His Successors

In 1924, the most distinguished German geographer of his time, Albrecht Penck, delivered an address before the Physical-Mathematical Section of the Prussian Academy of Sciences.[38] The title of this address was "Das Hauptproblem der physischen Anthropogeographie," and its theme was that mankind must now undertake a global appraisal of resources and determine how many people these resources could sustain.[39] This was a return to the type of investigation initiated by Ravenstein in 1890, then continued by Fircks in 1895 and by Ballod in 1912. These earlier efforts are summarized, but not before Penck castigated Ratzel for not undertaking such a study. Indeed, the first two words of the lecture are "Friedrich Ratzel." Ratzel, as we have seen, was engaged with other ideas, but it will be profitable to note briefly the main criticisms that Penck had of Ratzel's idea of the habitable world. After crediting him with opening new areas in the study of the geography of man and with the recognition that the present distribution of mankind is a result of historical development, and with the description of the traces and works of man, whether cities or ruins, all over the world's surface, Penck remarked, in effect, that Ratzel stopped at the point that should have been a beginning. Ratzel, he argued, recognized the influences of natural conditions on man, but did not provide "even a hint of the most important influence of man on the earth's surface, namely the

---

38. Albrecht Penck (1858–1945) was a geographer and geologist and served as professor of cultural geography at the Universities of Vienna and Berlin, among many other appointments. In Berlin, his work included the classification of climates and cultural geographer. He further developed Friedrich Ratzel's concept of *Lebensraum*.—Ed.

39. Albrecht Penck, *Das Hauptproblem der physischen Anthropogeographie* [The main problem of physical anthropogeography] (Verlag der Akademie der Wissenschaften, 1924).

creation of a cultural landscape," and especially of that great compulsion, which man imposes on his own nature, to nourish himself.[40]

Penck set about to remedy this situation by studying the "relationship between the surface of the earth and man ... created by the need of subsistence." The problem of subsistence and the land area (he neglected the sea because of the loss of precision involved) could be expressed in the formula $z = Lp/n$, where z was the number of people on Earth, L the land surface, p the average production of the Earth's area, and n the average food requirement. Total production is a function of climate, soil, and intensity of cultivation. World trade makes possible the existence of people remote from a productive area. It is a leveler: it permits high and unequal distributions of population, but it neither increases the total population nor increases the resources of the world. Of these factors, two, size and the productive energies of the world, may be regarded as constant "for the geographical present." Only the third factor, intensity of cultivation, is a variable. The number of people possible on Earth thus varies solely with the degree of advancement in cultivation, and even this cannot be extended beyond a certain maximum.

With this theoretical framework, Penck turned to the concrete problem of determining the maximum numbers the Earth would support. In the present state of knowledge, he regarded his own efforts as suggestive and as supplying a method rather than as a firm determination. His approach rested squarely on climate considerations. Climate influenced the amount of food that could be produced, and the amount of food determined the population size. He assumed, other things being equal, that a given maximum population density is related directly to a given climate, and chose a climatic classification—the 1918 classification of Köppen—as his point of departure.[41] An estimate based on this classification was possible because Hermann Wagner in 1921 had, by planimetric methods, computed the land and sea area of the eleven major climatic areas in the Köppen system as it existed in 1918. Later critics of this method forgot that Penck himself realized its pitfalls. For example, Ireland and Japan, both in the wet temperate regions, could not have the same potential density. Again, a hot wet climate of a primeval forest in lateritic soils would have a much lower potential capacity than a climate region with volcanic soils. For any great climatic region, the estimate must be lower than the highest reported, but this need not depend entirely on which. He then surveyed eleven regions with these conditions in mind and made estimates of the utmost

---

40. Albrecht Penck, *Die Spuren und Werke der Menschen an der Erdoberfläche*, vol. 2, pp. 261–372. [Unverified reference.—Ed.]

41. The Köppen climate classification system, developed by Russian German climatologist Wladimir Köppen in 1884, and revised by German climatologist Rudolf Geiger, is based on the concept that native vegetation is the best expression of climate.—Ed.

conceivable and the probable maximum number of the inhabitants for each of the eleven regions. The utmost conceivable population was about 16 billion, the probable maximum 7.689 billion, which, because of the error and roughness of the estimate, could be rounded off to 8 billion. The main result of this estimate, based on soil and climate, and as with similar efforts undertaken before, was that the tropics emerged as the area of the great human aggregations of the future, just as the temperate zones are today. The big question that then arose was whether the tropics would become the chief productive regions of the world and at the same time become the habitat of great populations, or whether, as Friedrich List thought, they should be the granaries for the temperate zones, which, according to Huntington, were the seats of highest energy and civilization.[42] The solution depended a great deal on whether trade or commerce of the future could retain its present significance, and on the possible exhaustion of coal and iron.

But regardless of speculations, one thing remained certain to Penck: human development unfolded in a limited space, and the number of people could not increase beyond a certain quantity. Moreover, although he felt that "the coming conquest of the tropics" would require a considerable effort of mankind, hunger, "the most powerful cause of instinctive human activity, would help accomplish it. Penck wrote in conclusion, however, that not all the land on Earth could be used for food. Clothing, wood, commercial buildings, and dwellings would demand an appreciable portion of the Earth's surface. The task of a geographically oriented world economy was to so choose those locations that the total production was a maximum one and that a minimum of energy would go into transport from the place of production to the place of consumption. He admitted that the world was still far from this ideal worldwide economy, and that a strong relationship had not been established between economics and agriculture even in the best-cultivated nations.

The nature of the habitable world, as described by Penck, was derived basically from ideas that came from the study of plant distributions and from observed increases in the population of the world. The Köppen climate classification system is based on amounts of rainfall and its distribution throughout the years, and the zones are delimited by establishing certain maximum and minimum temperatures. The actual determination of those boundaries in the Köppen system came from the observation of vegetation and its boundary changes with latitude and temperature. Köppen himself approached climatology from a background in botany. There is no population theory here except a general view that population is increasing and will eventually spread out over all the land that can grow edible food.

---

42. Georg Friedrich List (1789–1846) was a German economist who developed the concept of the National System of Innovation.—Ed.

Penck believed that geographers had a better approach to tackling such issues than the abstract Malthusianists. The present distribution of population in the habitable world, in his opinion, was the result of historical circumstances in which climate may have played some part, but the habitable world of the future was being molded imperiously by climate and population growth. The result, he argued, would be one of two alternatives: either the tropical lands would produce a great part of the world's food and simultaneously provide living room for large increases in population, or they would merely provide the food, with man depending on a highly developed world trade mechanism to preserve the present general distribution of the Earth's peoples. The future picture was a grim one in the sense that the cultivation of vast new stretches of the four-fifths of the globe still relatively unsettled could only be brought about through sacrifice, which only necessity would induce men to make. There was very little theory in Penck's conception. Like Ratzel's, it was a narrowing down of the idea of the habitable Earth to specific areas and regions, and concluding from this that with population growth, they would fill up until the whole Earth had been fully occupied. An appraisal based on climatic regions, even with all the reservations, even with the recognition of soil differences, however, represented a grouping together of too many dissimilarities, too many regions that were comparable only in their mutual adjustment to the Köppen classification. Penck regarded his work as provisional, and the tone is modest throughout, but it is doubtful that the conception had a real methodological value from the start. It is probably no better, and certainly no worse, than taking key countries such as the Philippines or Finland as multipliers to estimate the world's arable land, classified by soil types.

Penck's address, and subsequent papers on the same theme, inspired, especially among German geographers, other efforts to arrive at the total carrying capacity of the Earth. These efforts differed considerably. Two of the most painstaking attempts were made by Alois Fischer in 1925 and W. Holstein in 1937. Fischer in two exhaustive articles in the *Zeitschrift für Geopolitik* (1925) examined previous studies and offered one of his own, based not on the Köppen system but on the capacity of the various political divisions of the world, arriving at his total by simple summation.[43] His main criticism of the previous estimates was that one could not as-

---

43. The magazine *Zeitschrift für Geopolitik* was established in 1919 by General Karl Haushofer, who became professor of geography at the University of Munich. The concept of *Geopolitik* played a significant role in Nazi policy, and emphasized five key ideas: the organic state, a version of political collectivism; *Lebensraum*, now a justification for imperialism; autarchy, a form of economic protectionism; pan-regionalism, akin to that associated with the British Empire; and an emphasis on land power. Glacken does not provide the exact references to the cited article by Fischer and I have not been able to trace it.—Ed.

sume a figure that would be valid once and for all. Any attempt, he argued, would produce only an estimate of the carrying capacity that was possible with the fullest use of existing technological skills, for the technology of the future was unknown.

Fischer computed the area of each political subdivision of the world, its carrying capacity at internally controlled conditions, the highest possible density, also internally controlled, the population and its density in 1925, and the population as a percentage of the internally controlled carrying capacity. The results are familiar. Continental Australia at 5 percent of the capacity would support 120,000 people. Inner Africa at 5 percent would support a billion and a half, and Brazil at 4 percent would support 9 billion. China was already close to its capacity, and Germany and Great Britain had greatly exceeded it, necessitating imports. The three great areas of future settlement in this estimate would be interior Africa, followed by South and North America. Of the 6.2 billion total, then, interior Africa would have 1.5 billion; South America, 1.2 billion; North America, 500 million in the United States plus 250 million in Canada; and Central America, 0.8 billion, yielding a total of 3.5 billion. Fischer avoided estimating the time when this condition would come about, labeling such efforts devoid of scientific value and polemical. The only certainty, he claimed, was that the situation was more unfavorable for Europe and the white race in general.

Holstein based his analysis on agricultural economics and soil science, and drew on the work of Shantz and Marbut on African soils and vegetation.[44] The greater part of the memoir is devoted to a description and appraisal of twelve different regions of the world: oasis agriculture; regions whose main crop came in summer, the amount of the crop being determined by temperature; regions of a long growing time; the maize-growing regions; the regions corresponding to a Mediterranean climate; the special case of the Iberian peninsula and Asia Minor; regions permitting two crops per year from the same field; regions of single crops with high temperature requirements; regions of continual rainfall or of two rainy periods; the wet tropics, where crops and seeding can follow one another without interruption; tropical and subtropical regions whose elevation is a limiting factor for crops; and a last, miscellaneous category.

Holstein did not stress the importance of estimating the total possible population, arguing that "one can reasonably take the position that the size of the earth's population will always be determined by entirely different circumstances than by the probabilities of obtaining subsistence." There were three steps in Holstein's appraisal: (1) production per hectare for each region, expressed in calories; (2) deter-

---

44. H. L. R. Shantz, C. F. Marbut, and J. B. Kincer, *The Vegetation of Soils of Africa*, AGS Research Series 13 (New York: National Research Council and the American Geographical Society, 1923).

mination of the proportion of arable land and total land for each region, determined by planimetric methods; and (3) calculation of the possible population density by multiplying by a factor that indicates the best possible use that can be made of the land. This is assumed to be 40 percent for the whole world. He argued for a somewhat higher utilization of arable land for direct subsistence production "than is customary in Germany today."[45] A concrete example will make this method clear:

1. In a given region, one cultivated square kilometer will provide the calorie requirement for 361 persons.
2. In this region, arable land is 20 percent of the total area.
3. The number of persons this area could support per square kilometer is therefore 72 (20 percent of 361).
4. Assuming that the best possible use of the land is 40 percent, the possible density of the region would be 29 (40 percent of 72).

This calculation is repeated with varying calorie requirements by region and by the varying percentages of arable land until the potential population of the earth is computed

Comparisons between the estimates and the general implications of all three of them, Penck, Fischer, and Holstein, may be seen in the following summary table, with the numbers indicating acres per person:

	Penck	Fischer	Holstein
Earth	8.0	6.2	13.295
Europe	2.08	0.57	0.778
Asia	(for Europe and Asia)	1.70	2.8524
Africa	2.32	1.65	3.8025
Australia	0.48	0.28	0.4512
North America	1.12	0.8	1.3508
South America	2.0	1.2	4.0601

The optimism of these efforts arises out of the apparent extraordinary capacity of Africa and South America, to say nothing of the minimum of 280 million population for Australia and Oceania and the minimum of 800 million for North America. Much of the world, with great effort, it is true, is therefore still available for settlement, and in each estimate, the total population is at least triple the population of the world at the time the estimate was made. The most noteworthy general charac-

---

45. Again, Glacken does not provide a reference to Holstein's paper, and I have not been able to trace the origins of this quotation. However, there is a more recent book: Nicole Becker, *Tragfähigkeitsberechnungen von Ratzel, Penck, Fischer, Holstein und dem Club of Rome* (München: GRIN Verlag, 2004).—Ed.

teristics of these three estimates, which use varying methods in the detailed computation, is the assumption of a relatively unchanging physical environment. The soils, the arable land, are looked on as fixed and constant in area. The only future changes envisioned are improvements. The picture in every case is relatively the same: the future expansion of the human race, under stress of necessity, into all corners of the habitable Earth. No consideration is given to possible deterioration, and the whole question of tropical soils, on whose performance those estimates rest for fulfillment, are not treated at all, or only in passing. The whole subject of soil erosion is not considered, and neither is there any conception of the changes in plant and animal life that these changes might produce and that would not necessarily result in the steady improvement in land or its maintenance in its present condition. The optimism of these appraisals is completely dependent on these two ideas: the fertility of the uncultivated tropics and the continued performance or improvement of cultivated soils. Viewed in this light, the optimism rests on highly controversial premises. The corollary to all of this is that as population increases, it will spread, with hardship, but rationally, with the new settlements, when achieved, operating at a technical level consistent with the progress of scientific inquiry at that time.

Penck dignified the assessment of the Earth's capacity as a most important field of study and indispensable to the future of the human race. There is no debating this. But in his hands and in the work of those inspired by his work, too much was taken for granted, and the correlations between climate and soils were too elemental. It was really a reaction to the idea of man's conquest of nature in the sense of annihilation of obstacles, eliminating the disadvantages of space and distance through transport and trade. The price of the conquest would be high in human terms, but it would be a beneficial one.

KEY TRENDS IN NINETEENTH-CENTURY ENVIRONMENTALISM

This essay, which draws on the printed draft manuscripts, has two broad parts. The first part details five trends in the history of nineteenth-century environmental sciences: (1) the beginning of the scientific study of soils separate from the study of agriculture, (2) the emergence of scientific ecology, (3) increasing scientific exploration, (4) a coalescing faith in progress and the decline of the idea of the designed Earth (a theme that Glacken continues from *Traces*), and (5) new ideas about anthropogenic environmental change. In the second part of the essay Glacken sketches a comprehensive intellectual history of environmental determinism. The essay is particularly interesting for the sudden shift in rhetoric at the end, where Glacken makes the move from a careful chronicler to an unabashed critic not just of environmental deterministic ideas but of other similar metatheoretical generalizations as well.

Five Trends in Late Nineteenth-Century Environmental Thought

As one looks back from the vantage point of the second half of the twentieth century to the accomplishments of the second half of the nineteenth, one cannot escape a certain sense of wonder. Many trends apparent at that time formed a mold within which the problems of human cultures and their physical environments were to be considered. While these trends have been legion, some should be singled out for special consideration.[1]

---

[1]. All subheadings in this section, such as "Five Trends," have been introduced by the editor to aid the reader.—Ed.

Trend 1. Scientific Studies of Soils

The first of these trends was the beginning of the scientific study of soils. If one looks back at the history of agriculture and of soil study, one is struck by the fact that they have always been studied, empirically by common observations of yields, by rough classifications based on long observation, and more theoretically according to an underlying theory of chemistry or of geology. Much of early soil science was a legacy from Greek and Roman science and the theory of the four elements;[2] one sees evidence of this influence as late as the eighteenth century. Certain theories, such as Jethro Tull's idea that soil itself provided nourishment for plants,[3] competed with other theories holding that the real constituents of plant food in soils were in the humus of the soils. The humus theory derived in part from observations of the fertilizing qualities of decaying organic materials and in part from a belief in vitalism—that life could only come from and be supported by materials that had once been living themselves—which was strongly held up to the time of Justus von Liebig.[4] And although we must give full credit to Liebig's predecessors for pointing out the fatal flaws of the humus theory, it was his work, scientific and popular alike, that delivered the final blow to the theory. Even Liebig's theories, however, were restricted largely to arable soils: his main interest was in soils that could be cultivated. Broader horizons were possible when it was realized that soils in themselves, not merely as the agents of plant growth, were an appropriate subject for scientific inquiry. This new direction in scientific study we owe primarily to the Russian pedologists of the nineteenth century, Vasily Dokuchaev[5] and his school, and to a lesser

---

2. The four elements are air, fire, earth, and water and were thought to be the primordial origins of all life.—Ed.

3. Jethro Tull, *Horse-hoeing husbandry; or, An essay on the principles of vegetation and tillage: Designed to introduce a new method of culture, whereby the produce of land will be increased, and the usual expence lessened: Together with accurate descriptions and cuts of the instruments employed in it*, 4th ed. (London: A. Milar, 1762).

4. Justus von Liebig, *Chemistry in Its Application to Agriculture and Physiology* (London: Taylor and Walton, 1842). [The humus theory mentioned here held that the essential nutrients needed for plant growth were organic in nature, as opposed to chemical or mineral. Liebig argued against this theory and proposed instead that plants derived their nutrients from soil, carbon dioxide from the air, and hydrogen and oxygen from water, and in doing so he ushered in the chemical revolution in agriculture. Historians today, however, believe that Liebig was preceded in these claims by the German agronomist Carl Sprengel (1787–1859).—Ed.]

5. Vasilii Vasilevich Dokuchaev, *Russian Chernozem (Russkii Chernozem): Selected Works of V.V. Dokuchaev* (Jerusalem: Israel Program for Scientific Translations, 1967). [Vasily Dokuchaev (1846–1903) is regarded as the father of pedology, the study of soils in their natural contexts. A crater on Mars is named for him.—Ed.]

extent to the investigations, similar in character, of Eugene Hilgard[6] in the United States. Though there have been many refinements and changes in theory since that time, these investigators laid the groundwork for a consideration of soils as a product, broadly speaking, of climatic conditions. They argued that the processes of soil formation were related to the climatic regions of the Earth, thereby providing the basis for the modern theory of soils as products of slope, time, parent material, and climate. The modern theory of soils and their formation made possible further extensions into the study of tropical soils, so important in contemporary literature with respect to the expansion of populations into the tropics and the possibility of using tropical soils for agriculture. In the broadest sense, the major step was liberating the study of soils from the study of agriculture; the study of soils was no longer an adjunct of agriculture but a field of inquiry in its own right. Along with the developing field of climatology, the study of the world distribution of soils contributed to an understanding of wider environmental and physical areas on the Earth's surface, thus establishing the framework for a more inclusive and more general treatment of the principles of physical geography.

Trend 2. The Emergence of Scientific Ecology

The second major trend that appeared in the second half of the nineteenth century was a growing interest in ecology, an area of inquiry that in our own time has become important in all environmental study, especially that relating to human activities. I have argued elsewhere that we must look for the antecedents of modern ecological thinking in the old ideas of harmony and balance in nature, created and maintained by an all-perceptive creator.[7] But modern ecological theory undoubtedly owes its inspiration to the scientific discussions that centered on Darwinian theory, especially the notion of a web of life and the principle of the interrelatedness of things, a principle that had also been emphasized in the past, especially by

---

6. Eugene W. Hilgard, *Soils, Their Formation, Properties, Composition, and Relations to Climate and Plant Growth in the Humid and Arid Regions* (London: Macmillan, 1906). [Eugene Hilgard (1833–1916) was a German American chemist who had a distinguished academic career at the University of Michigan and the University of California, Berkeley. He also played a critical role in the American Civil War, during which he served as custodian of the University of Michigan's buildings, and in the Northern Transcontinental Survey, for which he directed the agricultural division.—Ed.]

7. In the original text, this sentence read: "It has been pointed out in previous chapters that we must look for the antecedents of modern ecological thinking in the old ideas of harmony and balance in nature created and maintained by an all-perceptive Creator." Alas, the chapters he refers to have apparently been lost. However, readers are advised to refer to his *Traces on the Rhodian Shore* for an elaboration of this particular point.—Ed.

Alexander von Humboldt in his plant geography. The newer Darwinism, with its emphasis on interrelatedness in nature (including as well certain Lamarckian ideas that Darwin had at first rejected), called attention first to the ecology of the individual organism (autecology). This trend is most noticeable in one of the earliest works on ecology as a separate science, that of Eugene Warming, published in the early years of the twentieth century.[8] Later, in the hands of such men as Prince Peter Kropotkin,[9] the idea of a struggle for existence was supplemented by ideas of cooperation and mutual aid in nature, and later still—perhaps the most important development from a geographic point of view—by an emphasis on relations among whole associations of plants and animals, which gave rise to theories about plant communities and plant successions. Classic studies such as those of Möbius on the oyster[10] and Lindeman on the lake as a microcosm[11] were superseded by works that showed the dynamic interrelationships between populations existing in associations and communities. Although the practitioners of this new science did not emphasize the role of human cultures in changing these interrelationships in nature, they again laid the groundwork for future investigation of problems of this kind. In fact, many of the earlier studies were so concerned with ecological and environmental relationships that they seemed at times to forget that human settlements and human populations existed that were ready to disrupt and interfere with these precious harmonies and balances observable in nature.

Trend 3. Scientific Exploration

A third general trend cannot be divorced from the other two, although it is more general in nature and less specific. This was the practice of scientific traveling in an age in which—at least compared to ours—political obstacles seemed to be the least of travelers' worries. A few examples will suffice. The *Challenger* expedition of 1872–76 gave us our first real understanding of the seas. Scientific travelers and

---

8. Eugenius Warming (1841–1924) was a Danish botanist and is considered one of the pioneers of the discipline of ecology.—Ed.

9. Petr Alekseevich Kropotkin, *Mutual Aid: A Factor of Evolution* (New York: McClure Philips, 1902). [Peter Kropotkin (1842–1921) was a Russian aristocrat and prince in Smolensk, a direct descendant of the Rurik dynasty, which preceded the Romanoffs. Kropotkin, a polymath, contributed to zoology, evolutionary theory, economics, geography, and philosophy, besides being a pioneer of the anarcho-communist tradition.—Ed.]

10. Karl August Möbius, *Ueber Austern- und Miesmuschelzucht und die Hebung derselben an den norddeutschen Küsten* (Berlin: Wiegandt und Hempel, 1870). [Karl Möbius (1825–1908) was a German zoologist. He opened the first German seawater aquarium in Hamburg in 1863 and was an expert on the ecology of oysters.—Ed.]

11. Raymond L. Lindeman, "The Trophic-Dynamic Aspect of Ecology," *Ecology* 23 (1942): 399–417. [Raymond Laurel Lindeman (1915–1942) was a pioneer in limnology and ecosystem ecology.—Ed.]

other interested men—sometimes local experts—wrote on the flora and fauna of widely scattered portions of the Earth's surface: China, India, Africa, Australia, and South America.[12] This was also the era of the scientific exploration of the American West. The colonial powers, interested in their lands and their investments, incidentally encouraged scientific studies of physical environments. Mining and gold rushes brought men to odd places on the Earth's surface, while international migration, perhaps the freest it has been in the history of the world, including our own times, promoted not only colonization but faith and optimism: a faith that trade and migration had solved the problem of food for peoples no longer dependent on the production of regions close to them. One gets the flavor of these journeys of exploration and migration by going through old geographic journals—Petermann's *Mitteilungen*[13] is probably the best and most inclusive.

Trend 4. Faith in Progress and the Decline of the Idea of the Designed Earth

Not the least of these trends was a growing faith in progress, not merely in improvement but in inevitable progress, whose dominance as a guiding thought in the nineteenth century has been so frequently and so correctly emphasized. The idea of progress was particularly important in interpreting many other trends. Problems in colonial areas, problems with the wanton destruction of nature in colonial lands (which bothered Friedrich and Brunhes alike), could be seen as less serious than they were, for they were merely temporary obstacles that could and would be overcome in the forward march of science, knowledge, morality, and social responsibility.[14] Despite the dissenters, the dominant note was faith in the machine, faith in the new science, and faith in technology, in much the same spirit as our contemporary

---

12. The H.M.S. *Challenger* expedition (1872–76) was a pioneering exercise in oceanography led by Charles Wyville Thomson and Captain George Nares. The scientists circumnavigated the world, traversed 70,000 nautical miles, and catalogued more than 4,700 previously unknown species.—Ed.

13. Augustus Heinrich Petermann, *Geographic Releases,* 1st ed. (Gotha, Germany: Perthes Verlag, 1855). [Augustus Heinrich Petermann (1822–78) was a German cartographer and a professor at the University of Göttingen who revived the *Geographical Yearbook,* published from 1850 to 1952, under the new title of *Mittheilungen aus Justus Perthes Geographischer Anstalt über wichtige neue Erforschungen auf dem Gesamtgebiet der Geographie von Dr. A. Petermann.* After his death in 1878 the journal was renamed *Dr. A. Petermann's Mittheilungen aus Justus Perthes' Geographischer Anstalt* and later, in 1938, *Petermanns Geographische Mitteilungen.*—Ed.]

14. Although Glacken does not provide an explicit reference, I believe that he is referring here to Ernst Friedrich and Jean Brunhes. Friedrich (1867–1937) was a German geographer who coined the term "*raubwirtschaft*" (plunder economy or destructive economy) to describe the destruction of land. Brunhes (1869–1930) was a French geographer, a student of Vidal de la Blache, a pioneer of the possibilist school of geography, which argued against environmental determinism.—Ed.

optimists view science as the answer to all human problems, including those of the growth of world populations.

With empirical and deductive modes of reasoning gaining dominance, the comforting philosophy of the designed world had disappeared almost completely from the world of science.[15] If scientists had not been the first to cut away the underpinnings of the view, they were potent in its destruction, but even in the new evolutionary theory there was room for emphasis, and the last paragraph of *The Origin of Species* shows far more faith in the doctrine of inevitable evolution and progress[16] than any pronouncement of Malthus ever did; indeed, one of the great spurs to the writing of the *Essay on Population* was the rejection of this idea as it had been expressed by Condorcet and Godwin.[17]

So the design idea did not survive in the scientific literature that had emerged by the middle of the nineteenth century.[18] Despite heroic efforts by believers in the old doctrines, such as the Reverend McCosh,[19] who generously expanded the design argument to include Darwinian theory as also a product of the creator's design, men turned their attention to nature itself; it was no longer necessary to study nature while always keeping one eye cocked for a sentence on the creator's forethought and wisdom or making sure that in the paths the scientist followed through nature he was but planting a smaller shoe in the large footprints left by the creator.

We all know that the doctrine of design had been subtly undermined long before Darwin's time, and that many of the scientists of previous centuries who believed in it, or said they did, nevertheless went about their work without special consideration of final causes and the argument from design. Yet it was far easier for these conceptions to linger in the life sciences—in botany, in natural history, in physiology—than in mathematics, physics, chemistry, or geology. And even if it is

---

15. In Glacken's original draft, this sentence started with: "It is true that. . . ."—Ed.

16. The last sentence of the *Origin* illustrates Glacken's point eloquently: "There is grandeur in this view of life, with its several powers, having been originally breathed by the Creator into a few forms or into one; and that, whilst this planet has gone circling on according to the fixed law of gravity, from so simple a beginning endless forms most beautiful and most wonderful have been, and are being evolved."—Ed.

17. William Godwin (1756–1836) published *An Enquiry Concerning Political Justice* in 1793. The Marquis de Condorcet (1743–94) published *The Future Progress of the Human Mind* in 1794. Both were utopian writers and believers in perfectibility and progress.—Ed.

18. The original text of this sentence was: "So, of the three ideas we have been discussing, only two survived the middle of the nineteenth century, survived that is in the scientific literature." The context for this sentence is, however, missing because the earlier chapters are lost.—Ed.

19. James McCosh, ed., *Method of Divine Government, Physical and Moral* (New York: Robert Carter and Bros., 1850). [James McCosh (1811–94) was a philosopher who attempted to reconcile Darwinist evolution with Christianity. He served for a brief period as the president of the precursor to Princeton University.—Ed.]

a false view to think that our conceptions of a benign nature as it existed on Earth, the product of a divine design, were suddenly shifted, like the scenery between acts of a play, to a harsh world of nature dominated by Malthusian principles, natural selection, and the struggle for existence, the evidence is nevertheless impressive for the strength of this idea, creation by design, in the nineteenth century. Students like James Fiske and William James show us its force, and a letter of Charles Lyell[20] clearly shows him hesitating to arouse the physico-theologists in the promulgation of his uniformitarian theory.[21]

Trend 5. Ideas about Anthropogenic Environmental Change

More obscure, more deeply embedded in the scientific and technical literature of the time, was the literature on environmental change caused by human agency. The European colonial expansion had brought along with it an interest in age-old methods of clearing, such as the fire clearing of the Africans, and in problems of deforestation, such as in India. The westward expansion of the United States not only produced its optimists and its theorists of the frontier, it also produced men who saw these movements in terms of the environments they altered. One might, with a great deal of truth, say that these late nineteenth-century conditions produced a milieu favorable to these two ideas, which, if not contradictory, at least emphasized two different aspects of the environment.[22] It was favorable, I think, for environmental theory, because of the deep study of history characteristic of the late nineteenth century and also the study of ethnography. The depth in which history was studied, the wide-ranging works on ancient and modern times, called forth, in some quarters at least, dissatisfaction with traditional historiography: many men, some not historians, were interested not only in history but in its geographic setting, in the geographic conditions that had made this history possible in the way it was enacted. Part of this trend we must undoubtedly ascribe to the historical and geographic writings of Carl Ritter: one can see in some of the famous historians of the late nineteenth century, such as Ernst Robert Curtius, Jules Michelet, and others, a growing insistence on considering not only the geographic background to but also the geographic influences on history.[23] Along with history, the accumu-

---

20. K. M. Lyell, ed., *Life, Letters, and Journals of Sir Charles Lyell*, 2 vols. (London: John Murray, 1881), 1:271.

21. James Fiske (1842–1901) was an American philosopher and exponent of Darwin's theories. William James (1842–1910) was an American philosopher and student of physiology and anatomy who, in developing his theory of pragmatism, applied Darwinian ideas to philosophy.—Ed.

22. The two ideas referred to are those of the optimists and theorists of the frontier, on the one hand, and those concerned with the environmental impacts of human actions on the other.—Ed.

23. Carl Ritter (1779–1859) occupied the first chair in geography at the University of Berlin and is considered one of the founders of the discipline of geography. Ernst Robert Curtius

lating materials on primitive peoples offered further opportunities for applying environmental ideas. It was not that every ethnographer ascribed cultural differences to the environment but that the collection of such a mass of materials gave ample opportunity for drawing superficial correlations between culture and environment. One could also, as has often been done, look on the expansion of peoples across whole continents, as happened in the United States, and the growing settlement and expansion in South America, Australia, and New Zealand, as geographically conditioned, an example being the role of the Appalachians in the westward expansion of America and their presumed influence in, if not preventing, at least channeling the course of the expansion. Wider knowledge of the world gave wider vistas, and for those to whom geography meant the study of influences and factors, the results of late nineteenth-century scholarship in history and ethnology represented an opportunity to investigate these influences not only in time but more thoroughly in their spatial distribution. Typical of this viewpoint was the introduction Lord Bryce wrote for the English translation of the multivolume set of Helmholt's *Universal History of the World*,[24] which had originally appeared in German. To Lord Bryce, the time had come to look at world history as a whole as it had been conditioned by the facts of geography.[25] For those to whom the traditional ideas of geography and geographic influence had come in the tradition of Hippocrates, Bodin, Montesquieu, Ritter, and Buckle, it seemed an unparalleled opportunity to give life to history, to rescue it from dynastic annals or from purely political narratives. Even Hegel, in the early part of the century, in his *Philosophy of History*, had seen its course as a manifestation of Spirit, but a Spirit that had been modified, conditioned, at least in a local way, by geographic circumstances.

An equally great opportunity was offered to those who were interested in other things, those who, while not necessarily interested in rebutting environmental theory, saw in science and technology, in the great international migrations and expansions, but another convincing, awe-inspiring manifestation of man's control over nature. It was the sort of confidence against which Miss Semple[26] had rebelled in her adaptation of the philosophy of Ratzel in her *Influences of Geographic Environment*. As she said in the preface, she was prompted to talk about these matters because man had been so "noisy" about his ability to control nature. This emphasis

(1886–1956) was a German literary scholar and philologist. Jules Michelet (1798–1874) was a French historian.—Ed.

24. Edward Gaylord Bourne, "Review of H. F. Helmholt, ed., *The History of the World: A Survey of Man's Record. With an Introductory Essay by The Right Hon. James Bryce*," *American Historical Review* 9, no. 1 (October 1903): 116–19 [New York: Dodd, Mead and Co., 1902, 4 vols.—Ed.]

25. Ibid.

26. Ellen Churchill Semple, *Influences of Geographic Environment* (New York: Henry Holt; London: Constable, 1911).

on the creativity of man, patent to everyone who would look, was only part of the story; man's control over nature was a sweeping phrase, an omnibus phrase meant to convey the great advances in theoretical science and their purposive application to conquer the physical obstacles offered by the environment: the building of railroads, harvesting and mining machinery, oceangoing liners, telegraph lines that annihilated distance. Transportation and communication improvements became the great tools by which the earlier restrictions posed by the environment were overcome in man's progressive control over nature. "Man's successes in his conflict with nature, won at an ever-increasing speed," said a 1913 reviewer of Lionel Lyde's *The Continent of Europe*,[27] "tend steadily to relieve in varying proportions the control, the inhibition, which relief and climate lay on human endeavor."[28]

These worldwide conditions also gave broad opportunities to those who were interested in the Earth itself, in the changes it was undergoing not only in the long-inhabited lands but also in the new lands being settled. Above all else it gave opportunities for comparisons, as Marsh's work so eloquently shows. The men of the late nineteenth century, looking behind the glibness of the phrase "man's control over nature," saw things that pleased them and some that displeased them, observations that, as they grew, were to lead, in the form of demands for legislation and conservation, to programs of forest care and preservation, the protection of nature, soil protection, and a vast array of other ideas associated with modern philosophies of conservation.

Environmental Theories

Speculations about the effects of climatic change on history are very old in the Western tradition: some of these have rested on theories independent of, some on theories dependent on, human agency. One of the oldest traditions is that concerning climatic charge as a result of the cutting down of trees, examples of which can be cited from Pliny's time. As early as 1799 Noah Webster[29] in an illuminating essay had investigated supposed evidence of climatic change based on materials in the classical and modern literature and had mercilessly criticized inferences drawn from insufficient data, often linked to memories of the old regarding climatic changes over short periods. Older theories of climatic change were more likely to

---

27. Lionel William Lyde, *The Continent of Europe* (London: Macmillan, 1930).
28. Glacken provided the following citation for this quotation: "(GJ 43, 1914, 70)." What exactly he was referring to is, however, unclear.—Ed.
29. Noah Webster, *Deforestation and Global Cooling: A New Theory for Disease*, Memoirs of the Connecticut Academy of Arts and Sciences (New Haven, CT: Oliver Steele, 1810). [Noah Webster was a medical doctor interested in the relationship between climate and human health; he researched and published on epidemic diseases such as yellow fever.—Ed.]

be based in beliefs that such changes had been brought about by human agency (an example being deforestation) than in beliefs about changes of climate independent of human agency.

Although one could write a whole book on the subject of climatic change and theories of climatic oscillation before the twentieth century, certain main theories and ideas do stand out because of the currency they attained and the influence they have exerted, in some cases to the present. Just when these modern theories about secular changes in climate, and about the desiccation of Eurasia and other parts of the world, got started is difficult to say. Historically they are associated with the never-ending interest in nomadic migrations in Central Asia and the problem of the migrations of peoples that at various times overwhelmed Europe, especially the barbarian migrations and the fall of the Roman Empire. Confronting one of the most dramatic events in the history of the West (as at least as it has been depicted by many historians), those who believed that Rome fell owing to migration rather than as a result of inner decay naturally sought for explanations such as overpopulation, the lust for adventure, and climatic change. In lectures to his students, Ferdinand von Richthofen[30] had advanced the idea of a postglacial increase in temperature that, with oscillatory reversals lasting to the present level of temperature, spread over the southern part of Central Asia, leading to desiccation and desert formation and forcing out the nomads living there.[31] The real impetus to the study of the modern climatic problem undoubtedly came, however, from two main sources, the arresting essay of Prince Kropotkin[32] on the desiccation of Eurasia[33] and the reports of the 1903 Pumpelly Turkestan expedition,[34] the latter largely through their

---

30. Ferdinand von Richthofen (1833–1905) was a German geographer who coined the term "Silk Road." He spent considerable time in China in the late nineteenth century, and served as professor of geology and geography at the Universities of Bonn and Leizig, and at Berlin's Friedrich Wilhelm University.—Ed.

31. Glacken provided the following reference for this: "Paul Rohrbach, Die. Gesch. d. Mensch. Lonigstein, 1929, 24. Rustow, I, 303, 70." However, I have not been able to locate the exact reference.—Ed.

32. Prince Pyotr Alexeyevich Kropotkin (1842–1921) was a Russian polymath and anarcho-communist. He was well known for his book, *Mutual Aid: A Factor of Evolution*.—Ed.

33. P. Kropotkin, "On the Desiccation of EurAsia and Some General Aspects of Desiccation, correspondence," *Geographical Journal* 43, no. 4 (1914): 451–58.

34. There were several expeditions involving the noted American geologist and Harvard professor during the period 1860–1905—geological explorations of China, Mongolia, and Japan in 1866; explorations in Turkestan in 1903, and in Turkestan and Anau in 1908. Glacken most likely is referring here to the explorations funded by the Carnegie Institution and undertaken from 1901 to 1905 in Turkestan, eastern Persia, and Sistan by Raphael Pumpelly; his son, Raphael Pumpelly II; Professor William M. Davis, another Harvard professor; and his assistant, Ellsworth Huntington.—Ed.

influence on the writings of Ellsworth Huntington, who as a young man participated in that expedition.[35]

Prince Kropotkin, who had done pioneer work on the orography of Asia and had himself explored Dzungaria and the Transbaikal from 1864 to 1869, advanced the theory that the continent of Eurasia was progressively drying up (although he made allowances for minor reverse fluctuations). This desiccation, Kropotkin thought, owed not to a change in rainfall but to conditions brought about by and as a consequence of the ice age. The immense masses of water tied up in ice were thawing and the glaciers' waters were gradually disappearing. It was not a question, therefore, of a lessening of rainfall; Kropotkin made it clear that if such a diminishment of rainfall had taken place, it was as a consequence, not a cause, of desiccation.[36]

Raphael Pumpelly,[37] who had made the celebrated explorations for the Carnegie Institute in Turkestan, especially at Anau, in 1903–4, had come to similar conclusions. Again it was the contrast between the past and the present, the buried cities discovered, and the vivid historical accounts of the nomadic movements into Europe that had called forth these questions—which indeed had puzzled men a century and a half before. I remember the illustration to an old edition of Volney's *Ruins of Empire*[38] in which the hero, lying under a palm tree, meditates sadly over the ruins about him, commanding and in the desert. Pumpelly saw connections among the buried cities of the Tarim, the decline in population and pasturage in Mongolia, the vanished sea of the Gobi Desert, the shrinking of the Aralo-Caspian undrained area, and the early and frequently repeated barbarian incursions into China and Europe. To Pumpelly and to the colleagues whose opinions he expressed, there was evidence of a "progressive desiccation throughout long climatic cycles in whose favorable extremes civilizations flourished which disappeared in the arid extremes."[39] Using the botanical evidence and the opinions of his col-

---

35. Ellsworth Huntington (1876–1947) was as a professor of geography at Yale. He is largely known for his work on climate determinism.—Ed.

36. Glacken does not provide a reference here, but it is instructive to read the following text: Prince Kropotkin, "The Desiccation of Eur-Asia," *Geographical Journal* 23, no. 6 (June 1904): 722–34.—Ed.

37. Raphael Pumpelly, William Morris Davis, and Ellsworth Huntington, *Explorations in Turkestan: With an Account of the Basin of Eastern Persia and Sistan. Expedition of 1903* (Washington, DC: Carnegie Institution of Washington, 1905).

38. C. F. Volney, *The Ruins, Or Meditation on the Revolutions of Empires: And the Law of Nature* (New York: Twentieth Century Publishing Co., 1890). [For the cover illustration mentioned here, see http://www.librarything.com/pic/186241.—Ed.]

39. Pumpelly, Davis, and Huntington, *Explorations in Turkestan*, xxxi.

league, Johann Ulrich Duerst,⁴⁰ Pumpelly saw in the appearance of wheat, barley, and domesticated animals (which Duerst had identified as first established in oases beyond the Caspian) in Babylonia, Egypt, during "the late Stone age in Europe," evidence not only of migrations from these areas made too arid for civilization but also of constructive reactions by the civilizations to which they had been brought to introductions that could no longer thrive in the centers of their origin.

Another theory of climatic change had been advanced by Rowland Thirlmere.[41] Thirlmere thought of climatic cycles as being of long duration, possibly measured in thousands of years. Over the years, however, both Kropotkin's and Thirlmere's theories were forgotten (they were revived briefly by J. W. Gregory in 1913 in a critical review, in which he dismissed the idea that there had been worldwide changes in climate in historical times),[42] but the theories of the Pumpelly expedition to Anau were, through the efforts of Ellsworth Huntington, to enjoy a reputation and an influence down to our own time, especially among historians. Toynbee's acceptance of the main outlines of Huntington's ideas is a case in point.

Huntington soon began writing at length on his favorite topic. In his first book, *The Pulse of Asia* (1907),[43] he argued for a progressive desiccation of Eurasia, but in the later *Civilization and Climate* (1915)[44] he switched to a pulsatory theory, which he thought could account for many of the vicissitudes in the rise and fall of empires. This theme remained with Huntington through his long productive life, and later studies of contemporary life allowed him to add breadth to what he had already advanced in historical depth. Through the study of various meteorological data (temperature, barometric pressure, seasonal changes in weather) he was able to his own satisfaction to identify high-energy and low-energy areas of the world's areas best suited for high civilization, from which the tropics were excluded. In the course of this research he returned to a tradition mentioned in the Hippocratic corpus, that seasonal change and irregularity, not monotony, stimulated men and nations. Although Huntington never wearied of explaining in his prefaces that he did

---

40. Johann Ulrich Duerst (1876–1950) was a Swiss zoologist and an expert on prehistoric cattle and horses. Among other things, he authored a noted book on the horse in Anau based on archeological findings. That book was also published by the Carnegie Institution of Washington.

41. Rowland Thirlmere, *Letters from Catalonia and Other Parts of Spain* (New York: Brentano's, 1905). [Thirlmere (John Walker, 1861–1932) was a writer and poet who also wrote a book, *The Clash of Empires*, on the threats German imperialism posed for Britain.—Ed.]

42. John Walter Gregory (1864–1932) was a British geologist and an expert on Australia and East Africa. Glacken provides the following citation: "GJ 43, 307," but it is unclear what exactly he was referring to.—Ed.

43. Ellsworth Huntington, *The Pulse of Asia: A Journey in Central Asia Illustrating the Geographic Basis of History* (New York: Houghton Mifflin, 1907).

44. Ellsworth Huntington, *Civilization and Climate* (New Haven, CT: Yale University Press, 1915).

not regard climate as the only determinant of the fate of present or historical civilizations, the arguments in the main body of his books inevitably made him the most famous and influential exponent of the deterministic point of view not only in the United States but in the international body of learning for at least a generation.

The excitement and interest with which these theories, especially the voluminous writings of Huntington, were received is well brought out in Huntington's paper, "The Burial of Olympia,"[45] which he read before the Royal Geographical Society in 1910, and in the discussion that followed. Though Huntington was modest in his claims and frank in his admissions of the gaps in knowledge that prevented full proofs of his theories, he nevertheless was doggedly determined to present the argument for the influence of climate on history. The immediate subject of his paper, the burial of the old grounds where the Olympic Games had been held, was lost in the wider implications of his theory. It is noteworthy, too, that the climatic theory brought into bold relief older theories of climatic change induced by human agency: the idea of man as a modifier of his environment thus showed its close relationship to ideas of environmental influence. It was the old question of deforestation and soil erosion in the Near East, in Greece, and in North Africa; the depredations of the goat—all supported by citations from the classical authors, including the celebrated passage from Plato's *Critias*.[46] The newer theories of climatic pulsation and desiccation thus must be seen not only as hypotheses of climatic change in their own right but as newer challenges to older ideas of deforestation that had been advanced for the eastern Mediterranean in scientific detail in the early part of the nineteenth century. In the discussion that followed the reading of Huntington's paper, the great names of Near Eastern history, David George Hogarth and J. L. Myres, and of Asian travel, Sven Anders Hedin and Sir Marc Aurel Stein, added their objections, which were based on historical data.[47] Although we cannot go into the many objections based on geography, geology, interpretations of climatic data, and the evidence from written history, it is important to stress that this new field

45. Ellsworth Huntington, "The Burial of Olympia: A Study in Climate and History," *Geographical Journal* 36 (1910): 657–75.

46. Glacken is likely referring to the following paragraph wherein Plato comments on deforestation in Attica: "What now remains compared with what then existed is like the skeleton of a sick man, all fat and soft earth having wasted away, and only the bare framework of the land being left ... there are some mountains which have nothing but food for bees, but they had trees not very long ago, and the rafters from those felled there to roof the largest buildings are still sound."—Ed.

47. David George Hogarth (1862–1927) was a British archeologist and Keeper of the Ashmolean; John Linton Myres (1869–1954) was a British archeologist and the Wyekeham Professor of Ancient History at Oxford. Sven Anders Hedin (1865–1952) was a Swedish geographer and explorer of Central Asia. Sir Marc Aurel Stein (1862–1943) was a Hungarian British archeologist of Central Asia. The discussion that Glacken is referring to here is not easily found.—Ed.

of study turned many people's thoughts away from ideas of human agency, operating locally and throughout time, as the cause of environmental change and toward causes—more general and often worldwide—that lay at the basis of world history, especially Eurasian history. In the hands of Huntington and many men of similar interests, however, the climatic hypothesis continued to be applied to areas other than the Old World, Huntington's applications of his theories to the ancient civilizations of the New World being a case in point.

These new interests were also the product of two different trends. The first was a growing interest in archeology and in the human past before written history—an interest in plant domestication and in the early distribution of the domesticated plants, in migrations and their effects on civilization, and in the combinations and fusions of peoples that had brought about the civilizations of the ancient East: the Babylonian and the Sumerian, the Egyptian, and the Chaldean. The second trend was a more contemporary one: an increasing interest in these areas as regions of possible settlement and colonization, the surveys and general exploration of which were part of a desire to determine, especially in Central Asia, new possibilities of human settlement in long-deserted areas. Huntington himself saw clearly the operation of these two trends, which lay at the basis of much theorizing and would lead to newer studies of the role of climatic change in human history. This newer study, it is readily apparent, differed from the older environmental determinism in that it was more dynamic than static, showing less interest in theories of the molding of character than in the consequences, apparent in migrations, wars, and the disappearance of civilizations, of long or short, continuous or pulsating, trends in the world's climate. Though much of the furor surrounding the work of Huntington has died down, and there has been a notable trend toward rejecting his theories as unsubstantiated, the interest in climatic change and its effect on history, cultures, and civilizations has continued to the present day. This interest, however, I think is much more modest in its aims, more local in its preoccupations. Scandinavian scholars, for example, have made use of climatic data in postulating important changes in the histories of their nations. A Norwegian historian has accepted the idea of climatic change as being at least partly responsible for the long gaps in Norwegian history previously attributed solely to the Black Death.[48] Newer Swedish works, cited by Manley, employ analysis of past climates in explaining the history of

---

48. Glacken does not provide a corroborative citation. However, it is instructive to read the following contemporary study correlating climate change with the Black Death: Sharon N. De-Witte and James W. Wood, "Selectivity of Black Death Mortality with Respect to Preexisting Health," *Proceedings of the National Academy of Sciences of the United States of America* 105, no. 5 (2008): 1436–41. See also Ole J. Benedictow, "The Black Death: The Greatest Catastrophe Ever," *History Today* 55, no. 3 (2005).—Ed.

settlement and vegetational change. Much of this new work has been based on pollen analysis and discoveries of plant residues whose occurrence at their present site is evidence of the existence of a different climate from the one now prevailing; carbon 14 dating has been used to pinpoint these with some accuracy, so that the climatic theories rest on surer evidence than those of Huntington's time. It is important to notice also that we are concerned here with climatic variations and changes associated not with the end of the glacial period but with historical time. Certainly much more can be said about them, and the field is still open. There will always be interest in showing such relationships between climatic change and human history, and it may well be that future research will induce more confidence in these fields than studies aiming for a worldwide application—like Huntington's—have enjoyed in the past. It is true, however, that many of these theories have been accepted uncritically, and the range seems to run from a complete acceptance of the main ideas of Huntington in such a work as Toynbee's[49] to the assertion of John Leighly[50] that all theories of climatic change of this kind may be ignored.

Meanwhile, without much attention to these theories, a traditional form of environmental determinism—often softened to theories of environmental influence—continued, seemingly along paths well marked out in the past. Each individual country in the West at least has had literature on the subject of the geographic factors or influences that molded it. Many of these works were written as contributions to the understanding of the particular country, but a very few gained recognition as significant contributions to international geographic thinking. An important exception was the work of Miss Ellen Churchill Semple, whose *Influences of Geographic Environment* (1911) and *The Geography of the Mediterranean Region* (1931) achieved an international reputation.[51] The first of these works, inspired by the example of Friedrich Ratzel, whose student she was, was more traditional in its outlook test than even Ratzel's first volume (of which it was an adaptation).[52]

---

49. Arnold J. Toynbee, *A Study of History*, 12 vols. (Oxford: Oxford University Press, 1946).

50. John Leighly (1895–1986) was professor and chair of geography at the University of California, Berkeley, and the first doctoral degree recipient from that department. He was a former student of Carl Sauer and, like many belonging to the Berkeley school started by Sauer, a strong critic of environmental determinism. See John Leighly, *Land and Life: A Selection from the Writings of Carl Ortwin Sauer* (Berkeley: University of California Press, 1963).—Ed.

51. Ellen Churchill Semple (1863–1932) was an American geographer and anthropologist best known for bringing to the attention of the Anglophone community the work of the German geographer Friedrich Ratzel. The works referred to here are *Influences of Geographic Environment: On the Basis of Ratzel's System of Anthropo-Geography* (1911) and *The Geography of the Mediterranean Region: Its Relation to Ancient History* (1931).—Ed.

52. Friedrich Ratzel (1844–1904) was a German geographer and anthropologist whose organic state theory and use of the term *Lebensraum* have been seen by some as ostensibly justifying imperial expansion. The "outlook test" refers to a central tenet of environmental determinism,

Miss Semple wrote in a manner that would have been understandable to the theorists of the nineteenth century and even to the eighteenth century. While not an avowed determinist, she felt that in the universal exultation over man's control over nature more lasting geographic truths were being neglected. Critical of some of Ratzel's unsubstantiated ideas, his accuracy, and his failure to document his materials properly, she avoided some of the most pregnant ideas in the German geographer's work. One can read *Influences of Geographic Environment* without learning much, if anything, of the immense impact of theories of evolution on geographic thought of the nineteenth and early twentieth centuries (especially in the work of her master), of the strong emphasis on migrations and their cultural consequences on human geography, or indeed of Ratzel's recognition of the power of man in changing his environment through clearing, burning tropical forests, and many other practices. Hers was a systematic treatment of influences—of rivers, islands, mountains, plains, and so forth—but one that made little use of the newer ideas that had revivified the thought of the late nineteenth century. Nevertheless, her work was very influential for her contemporary American students and for historians; her work on geographic influences in North American history, also conventional in outlook, stressed the physical barriers in the overcoming of which the people of the United States had achieved control over the varied environments of the continent.[53] In many ways, Miss Semple's work, important and interesting as it was, represented a point of view that was already passing and under attack; like Ritter's geographic ideas in the early nineteenth century, it represented more the close of an older era than the heralding of a new one in the realm of geographic ideas. Yet one must guard against the notion of single individuals and single books as closing or opening an era, and the closer one reads the works of even the most determined of the determinists the more one senses a note of caution, shared even by Huntington, about ascribing too much to the environment. Brigham, who, like Miss Semple, attempted to interpret the course of American history in a work published in the same year as Miss Semple's geographic interpretation of American

---

namely, that the physical and climatic characteristics of places have a strong impact on the psychological outlooks of the inhabitants of these places. One clichéd interpretation of the outlook approach is the claim that people in the tropics were less developed than those in higher latitudes because continuously warm weather meant they did not have to work as industriously as their northern counterparts to survive. Glacken is implying here that Semple made even stronger claims in this regard than Ratzel.—Ed.

53. Ellen Churchill Semple, *American History and Its Geographic Conditions* (Boston: Houghton Mifflin, 1903). [In his typescript, Glacken referred to this book as "Geographic Influences in American History," which is actually the title of Albert Brigham's book, cited next. I have modified the text to eliminate this error.—Ed.]

history, made explicit statements and warnings of the dangers of too great reliance on environmental factors.[54]

Even though the literature on environmental influences that has been produced so far in the twentieth century is very voluminous in English, French, and German, certain tendencies—inevitable because of the growth of knowledge in related fields—were making themselves felt, tendencies that sooner or later had to temper the enthusiasm of those who looked for environmental influences everywhere. Some of the most important of these must be mentioned here. General history writing and monographic histories, which accumulated so fast in the nineteenth century, made nonsense of some of the cruder analyses, which seemed to be little more than a combination of map study and history outlines. In the search for environmental influences, history could not be ignored. However, historical evidence could be embarrassingly vexing. For example, in his "Burial of Olympia," Huntington had expressed wonder that the evidence of the eastern side of Greece, drier than the western side, was difficult to reconcile (without postulating a climate-related drying out) with the greater population and the higher culture of eastern Greece.[55] Hogarth commented that when Greece was important, its most important trade route was actually on its eastern side, and that population centers had grown up in relation to it, with climate having little to do with it.[56] This is only an isolated example of the increasing urgency for students of geographic influences to inform themselves about historical facts and events that might modify or entirely negate a favorite correlation.[57]

Neither could the results of anthropological research be ignored. Ethnographers and anthropologists of course had had their day of interest in environmental theories (in common with most students of the social sciences), but simple correlations could not withstand the mounting evidence of cultural similarities and differences that required explanations of an entirely different order. In the United States, for example, under the critical leadership of such scholars as Alfred Kroeber and Robert Lowie, the correlations between environment and culture had to be abandoned if these proponents hoped to command a respectful hearing in circles beyond their own acquaintance.[58]

---

54. Glacken does not provide a citation, but he was no doubt referring to Albert Perry Brigham, *Geographic Influences in American History* (Boston: Ginn, 1903).—Ed.

55. Huntington, "The Burial of Olympia."

56. D. G. Hogarth, in "Discussion of 'The Burial of Olympia,' Huntington," *Geographical Journal* 36 (1910): 675.

57. This paragraph has been significantly edited for coherence.—Ed.

58. Alfred Louis Kroeber (1876–1960) and Robert Henry Lowie (1883–1957) were American cultural anthropologists who studied with Franz Boas at Columbia University and subsequently spent most of their careers at the University of California, Berkeley.—Ed.

The debates over environmental theories, however, became very heated in the discussion of the distribution of plants and animals.[59] An entry point to this debate is the reception of Victor Hehn's[60] book on crop and animal migration.[61] Through a study of philological materials, Hehn had inferred that many plants acclimatized and growing in the Mediterranean had been imported through human agency; they were not the plants that would necessarily have been there had these historical events not taken place. In rebuttal, Hehn's critics, though admitting to the existence of some exotics, maintained that the cultivated plants, like the natural vegetation, were responses to climatic conditions.[62] Hehn emphasized the importance of man as a carrier of domestic plants and animals in his wanderings, whereas his critics considered other climatic controls over plant distributions to be more important.[63] It is noteworthy here that historically, the classification of climates, in its modern scientific form, began with the observed distribution of plants: Wladimir Köppen's famous classification of climates, at least in the beginning, was inspired by the study of plant distributions made by Alphonse de Candolle.[64] There was a great deal of promise in this empirical approach: the more remote areas of the world physically, even the more accessible countries that had not developed scientific procedures for the study of weather or even for recording temperature or rainfall, potentially offered possibilities for those seeking to understand climate. Trained botanists, scientific travelers, and amateur natural historians, however, could travel in these regions and make observations, with the idea in the back of their mind that plants, unlike animals, could not move and had to adjust to the environmental conditions around them. With a critical mass of evidence gathered, even if it was of very uneven quality, it was possible to draw climatic vegetative zones on the Earth,

---

59. This transitional sentence was not present in Glacken's draft. This paragraph has, moreover, been significantly edited for coherence.—Ed.

60. Victor Hehn (1813–1890) was a German philologist and cultural historian.—Ed.

61. Victor Hehn, *Kulturpflanzen und Hausthiere in ihrem Übergang aus Asien nach Griechenland und Italien sowie das übrige Europa. Historisch-linguistische Skizzen* (Berlin: Gebrüder Borntraeger, 1870). The English translation was published as *Crops and Domestic Animals in Their Transition from Asia to Greece and Italy and the Rest of Europe: Historical and Linguistic Sketches* (Amsterdam: John Benjamins, 1976).—Ed.

62. Glacken does not provide any references as to who his critics were. However, for a broader context that defined Hehn's work, read O. Schrader, *Prehistoric Antiquities of the Aryan Peoples: A Manual of Comparative Philology and the Earliest Culture,* trans. Frank Byron Jevons (London: Charles Griffin, 1890). See esp. pp. 30–35.—Ed.

63. This paragraph has been significantly edited for coherence.—Ed.

64. This sentence has been edited for the paragraph to make more sense. Glacken provides a reference here: Alphonse de Candolle, *Origins of Cultivated Plants,* 2nd ed. (London: Kegan Paul, Trench and Co., 1886). Alphonse Louis Pierre Pyrame de Candolle (1806–93) was a French Swiss botanist best known for his contributions to botanical nomenclature.—Ed.

with necessary extrapolations for areas unknown or unsurveyed; and, assuming the value of plants as indicators of environmental conditions, especially of temperature and rainfall, it was possible to make a climatic classification. It was but completing the circle for those explaining the distribution of cultivated plants and the natural vegetation—using materials whose individual tolerances were known—to resort to climatic explanations.[65]

This type of thinking was particularly noticeable in writing from around the turn of the century, even in that of an avowed anti-environmentalist such as Lucien Febvre.[66] Undoubtedly, the growing science of plant ecology, with its emphasis, derived ultimately from Darwin, on adaptation to environmental conditions, gave respectability to this point of view. Environmentalist ideas could thus be pushed back from a place among human cultures and human societies, where their direct influence could not be sustained against the objections of other students of man, and moved to another level, that of plant and animal distributions, where there was much more evidence of their validity. So instead of being a direct influence on human cultures, environment—meaning, often, climate, temperature, and rainfall—exercised an indirect influence, modified by cultural and historical facts, on peoples and cultures. Though no one would deny the general validity of the approach, too rigid an insistence on the climate-plant correlation led to static views, downplayed the importance of plant migrations and the human artificial selection of plants grown in environments different from those in which they were held to have originated, and set aside the breeding of plants for special purposes (such as vernalization). It led indirectly to a kind of economic determinism based on geographic distributions that was to continue in our general textbooks down to the present.

Not all the criticisms of environmental theory came from those who were not concerned with it; geographers themselves had increasing misgivings. These misgivings required intellectual courage, for they attacked the fundamental aspirations of the discipline. In the latter part of the nineteenth century and certainly in many works of the twentieth, the term "scientific geography" did not mean a geography divorced from human concerns (such geomorphology and meteorology) but a geography devoted to ascertaining the effects and influence of environment not only in the past but in the present as well. To this day, even in universities, the task of the academic geographer is popularly held to be one of tracing and ascertaining the nature of geographic influence. Misgivings over environmental theory in rela-

65. This paragraph has been significantly edited for coherence.—Ed.
66. Lucien Febvre, *A Geographical Introduction to History* (London: Knopf, 1925). [Lucien Febvre (1878–1956) was a French historian known for being one of the founders of the Annales school of history.—Ed.]

tion to the geographic distributions of plants and animals, of course, reach far back in the nineteenth century; we can see them in the last quarter of the nineteenth century in such works as Alfred Kirchhoff's *Man and Earth*;[67] they were the very basis of the reconstructions of the French possibilist school and are fundamental in the approach of the theorists of this school, Paul Vidal de la Blache, Jean Bruhnes, and Lucien Febvre.[68] There was some impatience with the essential narrowness of the approach. Alfred Hettner[69] has written feelingly about the situation: the students of Ritter, he said, were lost in the blind alleys of seeking influences, for if one presupposed the existence of these influences, what else was there to do but to pile influence on influence, to go down one side of the street and up the other, to ransack history and contemporary life for further proofs and illustrations? What would one have as a result of such labors? A mass of materials on influences, something very traditional and yet unusable: the investigations did not lead to new vistas of learning and investigations; they merely led to conclusions based on the presuppositions. Hettner's further dissatisfactions led him to a broader view of geographic relationships, a point of view that in turn led him to a study of the historical aspects of the course of world history. Broader bases were needed than the older views had supplied, and in the nineteenth century these had come from scientific travelers, not from students of influences.

Another development, primarily of the twentieth century, but having its roots at least in part in the nineteenth, was the attempt, made largely by Sir Halford John Mackinder, to view the history of the world from the perspective of the relationships of large, critically situated areas or nation-states to one another and the changes in such relationships in historical times.[70] His primary concern at first was with the Eurasian continent: like Huntington and many others, he had been

---

67. Alfred Kirchhoff, *Man and Earth: The Reciprocal Relations and Influences of Man and His Environment* (London: G Routledge and Sons; New York: E. P. Dutton and Co., 1914). [Alfred Kirchhoff (1838–1907) was a German geographer and naturalist and held the chair in geography at the University of Halle.—Ed.]

68. Paul Vidal de la Blache (1845–1918) was a French geographer and one of the founders of the French school of geopolitics. Jean Brunhes (1869–1930) was also a French geographer. He held the first chair in human geography in the world (at Lausanne, in 1907). He also coined the term "social geography."—Ed.

69. Alfred Hettner, *Grundzüge der Länderkunde*, vol. 1 (Leipzig: Verlag von Otto Spamer, 1907), vol. 2 (Leipzig: Verlag und Druck von B. G. Teubner, 1932). [Alfred Hettner (1859–1941) was a German geographer known for his concept of *chorology*, the study of places and regions. —Ed.]

70. H. J. Mackinder, "The Geographical Pivot of History," *Geographical Journal* 23, no. 4 (April 1904): 421–37. [Halford John Mackinder (1861–1947) was an English geographer and founding father of the fields of geopolitics and geostrategy. He served as director of the London School of Economics.—Ed.]

struck by the obvious interconnections of Asian, especially Central Asian, history and European history. To him, European history could not be understood without knowledge of Eurasian relationships, the influence of the sea and land relationships, because of the peninsular character of modern Europe. In his first attempt at bringing these materials together, "The Geographical Pivot of History," published in the *Geographical Journal* in 1904, he tried to show these relationships as they had changed throughout historical time; his emphasis was on the European-Asian continent, but, unlike Huntington and Kropotkin, he was not concerned with theories of climatic change. He used the accepted explanations based on atmospheric circulation as it was observed at that time: the moisture of the seas covering Western Europe and providing its rainfall compared with the progressive dryness as one got into the interior of Asia, with the vast concentrations of human populations occurring on the rainy margins of the Eurasian continent, leaving the interior relatively uninhabited but nevertheless strategically placed in older times because of the mobility of the nomads, owing to the horse. In modern times, because of the possibilities of railroad building, it was able to exert a powerful influence on the destinies of the marginal peoples and ultimately on the whole world: it was a reassertion of the lasting importance, temporarily eclipsed by the naval power of Western Europe, of land power over sea power. Although we are not here concerned with a history of geographic ideas as they pertain to politics, international relations, and political geography, certain ideas in Mackinder's early essay point to a subtle fusion of many of the ideas we have been discussing.

The essay, written but eleven years after the publication of Turner's essay on the American frontier,[71] has much the same sense, applied to a world situation, of a fundamental turning point in history that was marked not by the closing of the frontiers as such but by the final parceling out of the world among its nation-states, though Mackinder did not deny the possibility of further exploration. Mackinder wrote: "The missionary, the conqueror, the farmer, the miner, and, of late, the engineer, have followed so closely in the traveller's footsteps that the world, in its remoter borders, has hardly been revealed before we must chronicle its virtually complete political appropriation."[72] It was a theme, without the Malthusian additions, strikingly similar to that of Walter Prescott Webb in the 1950s.[73] In pointing out these similarities between Turner and Mackinder I am not suggesting any direct relationship between the two, merely that both were impressed with the fact

---

71. Frederick Jackson Turner, *The Frontier in American History* (1893; repr., New York: Henry Holt, 1921).
72. Mackinder, "The Geographical Pivot of History," 421.—Eds.
73. Walter Prescott Webb, *The Great Frontier* (Boston: Houghton Mifflin, 1952). [Walter Prescott Webb (1888–1963) was a U.S. historian of the American West.—Ed.]

that there was a new closed system (in each of their spheres of interest), and that the future destiny of mankind was now bound up with a world thoroughly staked out and whose national boundaries were known, and that exploration in the future would take the form of more intensive survey and "philosophical synthesis."[74]

It was these very circumstances that came into being shortly after 1900, the closed system of the world composed of nation-states, that in Mackinder's view offered the opportunity for the first time "to attempt, with some degree of completeness, a correlation between the larger geographical and the larger historical generalizations."[75] In this attempt, he dismissed investigation of "the influence of this or that kind of feature" (an abandonment of the traditional approach to the study of environmental influences) in favor of a view that showed "human history as part of the life of the world organism." In this structure, there was the idea of an overall limitation of the environment, not in the Malthusian sense but in the sense that "Man and not nature initiates, but nature in large measure controls." He nevertheless saw that within the limits of this overall environmental control (never expressly defined), human modifications of the landscape had produced modifications that added complexities to the simple antithesis of human history and the limiting effects of the environment. The effect of these changes was in some measure to relieve the strict control of the environment as it existed in conditions of pristine nature. "In all the west of Russia, save in the far north, the clearing of the forests, the drainage of the marshes, and the tillage of the steppes have recently averaged the character of the landscape, and in large measure obliterated a distinction which was formerly very coercive of humanity."[76]

In reading this essay, as in reading the writings of Huntington and Kropotkin, it is impossible not to be struck by the stimulus given to these men's thought by the successive migrations and incursions from the Asian interior; these migrations, literally spreading over thousands of years, but particularly dramatic in the case of the Goths, the Huns, and the Mongols, seemed to be the real incentive for the study of the physical conditions and the geographic relationships that produced them, or at least had an influence on their force and direction. Much of the history of modern Europe, Mackinder felt, could be written as the result of the stimulus and the impact of the migrations and incursions from the east, combined with the pressures, predominantly from the north, expressed in the Viking period.

In Mackinder's *Democratic Ideals and Reality*,[77] these old interpretations are not

---

74. This paragraph has been significantly edited for coherence.—Ed.
75. Mackinder, "The Geographical Pivot of History," 422.—Ed.
76. Ibid., 432.—Ed.
77. H. J. Mackinder, *Democratic Ideals and Reality: A Study in the Politics of Reconstruction* (New York: Henry Holt, 1919).

lost but are subordinated to the new realities coming into being as a result of World War I. Mackinder still retained his ideas of the world as a closed system; society was a going concern, and the new political realities, shown by previous history, gave renewed importance to the heartland of Eurasia and the heartland of Europe (Germany), and invoked the central importance of the Eurasian continent as the world island, surrounded by the lesser islands of North and South America and Australia. While this is not the place to analyze these arguments, which are extremely well known because of the immense revival of interest in them during World War II and because of their use by the German geopolitical writers, the real point in its widest sense is that Mackinder retained his earlier view of a broader environmental influence, not the minutiae of individual influences, and the broader generalizations therefore possible in considering world history and world geography. Despite the great amount of criticism Mackinder's thesis received, his work had the merit of seeing Eurasian history as a whole. He thought he was the first to do so, although, as Taggard pointed out in an early appreciation of Mackinder, he had a predecessor in Klaproth's *Tableaux historiques de la Asie*.[78]

Neither can we neglect the implications of Turner's frontier theory in this survey of environmental theories. As has been argued in detail by his many critics, Turner was not the first who pointed to the significance of the frontier, and many other writers may have had a more realistic appreciation of its significance. The theory has taken on additional significance because it has been extended to the whole world. Even the New World origins of the frontier theory have been questioned, for there are striking passages in Hegel's *Philosophy of History* contrasting Europe and North America. A real state, a real government, Hegel said, arises with class distinctions, with extremes of wealth and poverty, and when people can no longer satisfy their desires in the way they had. America had not experienced these problems because of the outlet of colonization in its own territories. With this colonization,

> the chief source of discontent is removed, and the continuation of the existing civil condition is guaranteed. Europe and America cannot be compared, even considering the European emigrations that have taken place.... Had the woods of Germany been in existence, the French Revolution would not have occurred. North America will be comparable with Europe only after the immeasurable space which that country pre-

---

78. Heinrich Julius Klaporth, *Tableaux historiques de l'Asie: Depuis la monarchie de Cyrus jusqu'à nos jours, accompagnés de recherches historiques et ethnographiques sur cette partie du monde* (Paris: de l'Imprimerie de Lachevardière Fils, 1826). It is unclear who the Taggard Glacken refers to is. I suspect Glacken intended to refer to Teggart instead. Klaporth (1783–1835) was a German linguist, philologist, and explorer, particularly known for his work on the Caucasus.—Ed.

sents to its inhabitants shall have been occupied, and the members of the political body shall have begun to be pressed back on each other. North America is still in the condition of having land to begin to cultivate. Only when, as in Europe, the direct increase of agriculturalists is checked will the inhabitants, instead of pressing outward to occupy the fields, press inward upon each other—pursuing town occupations, and trading with their fellow citizens; and so form a compact system of civil society, and require an organized state.[79]

It is true that there is missing from this analysis some of the distinctive aspects of Turner's frontier hypothesis—the lectures were first delivered in 1822–23. Turner's theory is tacked on to the formal statement of the closing of the frontier that appeared in the census report of 1890; it thus marked the close of an era that had seen certain distinctive processes in action that had molded in the American environment the American character, among them an application of the biological analogy (Turner calls this the "germ theory of the state") and analogies derived from geology, in which he was interested; an example is his expressive use of the analogy of the moraine in describing the forward-moving frontier. The frontier theory really is a part of the history of American theory and historiography; from a world point of view, its interest lies more in the stimulus it has given to other areas of research. An interesting example of this is the work of Walter Prescott Webb, who saw a closing of the world frontier by applying conclusions drawn from world population statistics (indicating the rapid rise in the population of the Earth as a whole) to the idea that the period from the age of discovery to the end of World War I represented a period of the expanding world frontier, closed by the era of immigration restriction, restrictions in travel, and a ratcheting up of internal problems, much as Hegel and Turner had pointed out for the United States.[80] All these points of view are, in a very real sense, environmentalistic in nature: they constantly remind their readers of the limitations of nature, despite man's creative capacities, limitations that may be purely physical or may be indirectly physical, as in the modern era's recasting of Malthusian theory, itself fundamentally an environmental theory since the ability of the Earth as a whole to provide sustenance for its people is always lower than the actual potential population.

The attacks on this point of view have come from those who have questioned the validity of "closed space" thinking; the very word "frontier" has been questioned and instead has been given a metaphorical meaning: if the actual physical frontiers have been reached, there is no limit to the new "frontiers" of knowledge,

79. G. W. F. Hegel, *Philosophy of History*, trans. J. Sibree (New York: American Home Library Co., 1902 [1857]), 141–42.
80. Walter Prescott Webb, *The Great Frontier* (Boston: Houghton Mifflin, 1952).

human creativity, and so on. Just as in the case of the pessimism of those who have objected to the destructiveness of mankind of the environment, so with those who have pointed to environmental limitations (independent of what man may do to the environment) imposed by the nature of the world itself. The distinction is an important one. Although it would be foolish to insist that these two trends have existed without any interaction between them, they do represent two distinct and readily identifiable trends. The one we are discussing here I think owes its inspiration to observations like those made by Robert Wallace[81] in the eighteenth century in his discussions of the absolute limits to population growth. Malthusian theory, more explicit, more widely recognized, and international in its influence, had the same effect: the insistence on the very real limitations offered by the Earth itself to human advancement and well-being. One can see this type of thinking applied to the social scene in many ways, among them the so-called man-land ratio. Sumner and Keller, for example, went so far as to see the man-land ratio (fundamentally drawn from Malthusian theory) as being the very basis of the formation of societal customs and folkways, the argument being that the customs surrounding the relation of population to food supply in various countries gradually solidified into a body of tradition and folkways.[82] Although Malthusian ideas are absent from both Turner and Mackinder, theirs was a different way of calling attention to the same problem: with Turner, it was the problem of American history; with Mackinder, the problem of world historical and geographic relationships imposed on a world well bounded and parceled out. The pessimism of this literature would have been far deeper had thinkers of this type informed themselves adequately of the growing literature on actual environmental destruction by human societies.

In many ways, the attempts of Mackinder and of the students of Turner, who extended Turner's theories to cover the whole world, were an improvement over the directional philosophies of the geographic course of history. In Hegel's and in Ritter's time, there was a preoccupation with the westward march of civilization

---

81. Robert Wallace, *Various prospects of mankind, nature and providence: 1761. To which is added Ignorance and superstition, a source of violence and cruelty*, a sermon preached in the High Church of Edinburgh, January 6, 1746 (London: Printed for A. Millar, 1761; New York: Augustus M. Kelley, 1969, reprint of 1746 edition). Robert Wallace (1697–1771) was a Scottish clergyman, proto-economist, and demographer who is said to have influenced Malthus.—Eds.

82. The "man-land ratio" refers to the amount of land available per capita. William Graham Sumner (1840–1910) was the first professor of sociology at Yale College. He was a noted anti-imperialist and is said to have coined the term "ethnocentrism." Albert Galloway Keller (1874–1956) was also a Yale sociologist and succeeded Sumner as professor in 1907. Sumner and Keller wrote the classic *The Science of Society* (New Haven, CT: Yale University Press; London: Oxford University Press, 1927). Glacken cites Albert Galloway Keller, ed., *Earth-Hunger and Other Essays* (New Haven, CT: Yale University Press, 1913).—Eds.

("*ex oriente lux*"), resting, I think, largely on the belief in the origin of civilization in China. Hegel wrote that "with the Empire of China . . . History has to begin, for it is the oldest, as far as history gives us any information; and its principle has such substantiality, that for the empire in question it is at once the oldest and the newest."[83] The interest in Sanskrit and its origin, the interest in the home of the Aryans (which occupied a large part of the historical and philological activity of the nineteenth century), gave added impetus to the belief in the geographic march of history. The term "geographic march of history," which is an old one, was best defined by Guyot, a student of Ritter, as meaning that "the civilizations representing the highest degree of culture ever attained by man, at the different periods of his history, do not succeed each other in the same places, but pass from one country to another, from one continent to another, following a certain order. This order may be called the *geographical march of history*."[84]

The Aryan controversy, however, was not of much help, as it was so mixed up with conjecture and chauvinism. Some favorite choices for the Indo-European home were India, Central Asia, the highlands of western Asia, and the regions between the Himalayas and the Caucasus. On the other hand, there were objectors to the Asiatic origin theory: to De Halloy, the Aryans of Persia and India were conquerors from Europe; Boucher de Perthes saw the presence of man in Gaul at the beginning of the Quaternary age; Benfly saw in the immemorial residence of peoples of Europe a refutation of the Asian origin theory; Geiger in 1871 found it in central and western Germany; Cano thought it was to be found between the Urals and the Atlantic. In 1874, however, Virchow revived the theory of the Asian origin of the Aryans, returning to the point of view prevalent in the early part of the century, as expressed by Pott: "la marche de la civilisation a . . . suivi celle du soleil," or "the march of civilization follows that of the Sun." For those who accepted the Asian origin theory, the geographic march was plain enough: from Asia to Eastern or Western (depending on the point of view) Europe, and then on to the New World.[85]

---

83. G. W. F. Hegel, *Philosophy of History*, trans. J. Sibree (New York: American Home Library Co. 1902 [1857]), 176.

84. Interestingly, Glacken's text here cites Jules Guyot (1807–72), a French physician and agronomist. However, he is likely referring to Arnold Guyot (1807–84), a Swiss American geologist and geographer and author of, among other works, *Earth and Man: Lectures on Comparative Physical Geography in Its Relation to the History of Mankind*, trans. Cornelius Conway Felton (Boston: Gould, Kendall, and Lincoln, 1849).—Ed.

85. Rudolph Carl Virchow (1821–1902) was a German doctor, anthropologist, and prehistorian who is renowned for his contributions to modern pathology. He also was an early pioneer of craniometry, and ostensibly argued against theories of superiority of any given race over another, thereby debunking scientific racism. I have not been able to locate his writings on the Asian origin of the Aryans.—Ed.

Modern trends in scholarship, especially scholarship on prehistoric and modern times, have, however, made theories of this type very difficult to support. They were more tenable at a time when it seemed that the transplanted European culture in the New World was the logical successor to Europe, and they tended to ignore almost completely (except as certain events touched on Western history) the whole of Asian history. With the contemporary interest in development all over the world, with the plans of the Chinese government for the development of Central Asia and Russia's plans to develop the Far North, it would seem that what civilization is left is going without any inevitability whatsoever in all the directions of the compass. It was perhaps inevitable that some philosophies of history would emerge out of the obvious modern predominance of Europe and the quick expansion of the United States in the nineteenth century in a period when it seemed that other parts of the world were relatively stationary in their cultural inertia. For the theory seemed to rest on the belief in a long period of stagnation and inertia, following initial successes, in the history of the Eastern countries, with the superior science and the technological progress of the West providing the chief instrumentalities for the further extensions of civilization.

What is the importance of environmentalistic ideas today; how are they regarded, and what role do they play in the study of cultural history and of contemporary society? This is a very difficult question to answer because questions of environmental influence of this sort appear in a bewildering variety of works: in histories of art and of civilization, in regional and systematic geographies, in popular works, in interpretations of history, in studies of acclimatization. One could compile an imposing list of references that would indicate the great vitality of the question, but often such statements are confessions of personal belief, assertions made without proof, or affirmations based on personal experience. Before attempting to answer the question specifically, it is appropriate to summarize the confusions that have arisen almost entirely out of the history of the subject. Contemporary discussions mix up theories of environmental influence because they were never cleared up historically, or, if they were, the elucidations were forgotten. The original error, which appears in the Hippocratic writings, was the assumption that environmental influences on individuals, even if they seemed likely of proof, could by extension be applied to whole peoples. This elementary fallacy is conspicuous in much of the literature of environmentalism down to our own times. It seems obvious that a study of the influence of the factors molding a culture cannot be the same as one dealing with environmental effects on the human body or the human mind. One study belongs to anthropology, ethnology, sociology, human geography, and social psychology, the other primarily to physiology and psychology. This fallacy has been repeatedly pointed out by many writers, most clearly by

Herder in the eighteenth century. Throughout their long long history, the primary concern with environmental theories, as I interpret their history, has been with their value as tools for explaining cultural differences. This was certainly true of the Hippocratic doctrine and of other ancient speculations derived from it; it was true of the uses of environmentalistic notions in the eighteenth century, most reserved in Herder, explicit in Montesquieu. Even while confusing the individual and the society, the ultimate aim was to explain why one people were different from another, or, in modern terminology, to explain the effect of the whole environment on culture. All through its history, however, the question of cultural differences has been intertwined with other environmentalistic ideas: the effects on individuals, the effects on man through the agency of environmental influences on plants. We see today that many environmental theories are not theories at all but a potpourri of speculations on a wide variety of subjects. I am convinced that it is a meaningless quest to seek to determine environmental influences on plants. A whole culture: how is one to define the environment (itself a catchall term), and, if one is successful in discovering the authentic elements, how can one analyze their supporting or contradictory effects on a whole culture, itself the product of a long history? Environmental influences may be local in their application, well defined in their time and place, and these may legitimately be studied. It is a legitimate question to ask to what extent did the relief of Greece affect the trade routes of the country; perhaps "affect" is too strong a word, even; perhaps we should content ourselves with stating the known trade routes and the passes and valleys they went through, noting changes throughout time, changes that may have had nothing to do with the Greek environment but with world trade of the time, and how new roads, new passes, and new valleys came into prominence with new conditions. Questions concerning environmental influences on individuals or on plants and animals are not legitimate questions for a social scientist or a general interpreter to investigate unless he is prepared and competent to deal with them as a trained biologist, physiologist, or plant physiologist. It is not environmentalism to say that wheat has certain climatic tolerances; it is environmentalism to equate the distribution of wheat with climate and to assess the role of wheat eaters in the history of the world. It is not environmentalism to show that for the most part, the Norwegians have established their farms on terraces in deposits left by the seas and not on the stony mountaintops. These are matters of demonstrable fact. But it would be environmentalism to explain the history of the country by this circumstance. These confusions have been the curse of environmentalistic ideas. The difficulties arise not out of studying obvious environmental situations (like the aridity of the lee side) but in making cultural correlations without regard to historical and ethnographic data.

Even in our most recent discussions of environmental influences, these confusions are apparent. There has never been, there is not now, and there will never be a science of environmental factors. Through historical study, it may be possible to prove or to show the strong likelihood of a climatic change in Sweden having an influence on the population growth of the country, as Gustaf Utterström[86] has done; it may be possible to show shifts in trade and in economic position because of good and bad crop years—climatically induced—that affect local production. But these are totally different questions; they cannot all be grouped, like headings in a child's outline of his English composition, under environmental influences. Each is a problem of its own. Utterström's analysis is as good as his data, which may prove entirely, may show a strong likelihood, or may show a trend merely in predicting a connection between climatic change and population history. The evidence for proving this kind of relationship will be entirely different from that which seeks to show a correlation between human energies and climate, or that which seeks to show indirect influences through climatic influences on plants. Sometimes they rest on no evidence at all, merely contemporaneity or conjuncture, the causal relationship being inferred.

In the long history of environmental influences, few ideas have had as many funerals and as many active and honorary pallbearers. And few have had as many resurrections. One has the feeling that Strabo[87] felt he had finished Poseidonios off for good when he said that climate was not all, that education and institutions played their part; one has the feeling that Voltaire and Hume felt they had finished off Montesquieu for the same reason, and that Herder felt he had a synthesis that at least would prevent returning to old but honored and outmoded ideas. One has the feeling that Lefebvre in his bold attack on the Church fathers of the environmentalistic theory felt he had made any further attack unnecessary. Like Buckle, Spengler, and, later, Toynbee, the works of Huntington produced a large body of critical comment of their own. And yet why do such theories persist? Is it because men do not read criticisms? Do critics have a shorter life than synthesizers? Or is it that when all the objections are raised and considered, men still mutter under their breath, "And still it movest"? Why is it that geographers continue to turn to the questions of environmentalism and possibilism as questions that can be studied in their own right?

To the first three questions we can give a qualified yes answer. A bound work, if it becomes famous, is certainly better known than the criticisms of it scattered in a variety of places. They have a better chance of survival if they are gathered to-

---

86. Gustaf Utterström, "Climatic Fluctuations and Population Problems in Early Modern History," *Scandinavian Economic History Review* 3, no. 1 (1955).

87. Strabo, *Geographica* (23 AD), in *The Geography of Strabo*, trans. H. C. Hamilton and W. Falconer, 3 vols. (London: Henry G. Bohn, John Childs and Son, 1848).

gether, as were the criticisms of Buckle, Spengler, and of Toynbee. In studying the history of ideas, students are more likely to read and emphasize the positive affirmations made by the leading proponents of the idea, not the demurrers that gnaw away at it. These are insufficient explanations, however, and I think we must look elsewhere for the survival of these discussions to the present. They may be revived by influential writers and made to serve important roles in an interesting interpretation of history, as happened with the challenge and response idea of Toynbee in his *A Study of History*. They may be reasserted by men of acknowledged stature in a field, as in Griffith Taylor's unabashed environmentalistic ideas. The main reason, I think, however, is that they have an almost irascible appeal: living on an Earth with obvious contrasts in relief, climate, soils, and living conditions, even sophisticated scholars unacquainted with the long history of environmentalistic ideas may approach the subject with a sense of discovery, especially if they have an interest in geography: to many of these men, geography means geographic factors and influences.

There is little doubt that Toynbee's *A Study of History* revived interest in environmental theories and gave an even wider publicity to ideas already published, in particular the notions of Huntington on climatic change.[88] Toynbee's handling of these environmentalistic questions in itself is of interest. In the introductory volumes, he examines the various theories offered for the genesis of civilization. He considers the theory of the milieu and rejects it; his rejection itself is interesting because of the representative thinkers he chooses: Hippocrates and Desmolins, one writing in the fifth century BC, the other in the middle of the nineteenth century. Hippocrates is a logical choice because the essay is the first on the subject we have recorded in the Western world; as a classicist and a translator of Hippocrates, Toynbee could not be expected to ignore him. It is a surprise that Hippocrates could be taken seriously, in a work written in 1934, as a considered theory worth examining for its own sake rather than a historical document of inestimable worth. The choice of Desmolins, who wrote a three-volume work on how the route determined the social progress of civilization, is even more surprising. Severely, even men such as Ratzel criticized Desmolins, who had attained a certain notoriety in his day, themselves accused of environmentalism. The great names in the theory, with the exception of Hippocrates, are omitted. The works of Huntington complete the trio. Although Toynbee's rejection of environmentalist ideas must be taken in good

---

88. Arnold Joseph Toynbee (1889–1975) was a British historian who wrote a twelve-volume analysis of the rise and fall of civilizations, titled *A Study of History* (Oxford: Oxford University Press, 1934).—Ed.

faith, there is no doubt whatsoever that they enter in an important manner into his theory of the genesis of civilizations. Although environmental theories play a small part in Toynbee's overall philosophy of history, this part is a crucial one. Huntington's ideas are applied to specific historical circumstances: for example, his theory of desiccation is applied to the history of Eurasia and to the migrations of the nomads; Huntington is also called on as a support in the analysis of the Mayan civilization of the New World. More important is Toynbee's use of environmental ideas in the challenge and response theory, which is a crucial part of his analysis of the genesis of civilizations. In early periods of history, and among lower cultures, this challenge and response is conceived of in physical terms. Perhaps the most important and most general idea here is that of Herodotus (and Herodotus may be part of the Hippocratic tradition here): a real civilization, to be lasting, has demands made of it by the environment; ease is inimical to civilization. Toynbee's term "the stimulus of hard environments" could be used to describe the general impression coming from Herodotus's remark with which he closes his histories. The theory, however, is much more refined and sophisticated than this, for Toynbee believed that certain environments may be so hard, so uninviting, so difficult, that the challenge is too great to evoke a response. There is an optimum ground for the challenge and the response. It is the successful response of a civilization to the individual environmental challenge that prepares a civilization to mount responses to other challenges that are, however, of an entirely different order.

But even the environmental challenges are widely different among different peoples at different times. Among the unrelated civilizations (coming out of the yin-yang state of primitive society), desiccation may have been the challenge, as in the Egyptaic, Sumeric, and Minoan civilizations; marsh, bush, and floods plus temperature variation may have been challenges to the Sinic civilizations, and the tropical forest to the Mayan civilization. In Toynbee's system it is the challenge of the pristine environment that is uppermost, although he does recognize (as in the case of Ceylon) the effects of deforestation on the decline of a civilization. In one of his most controversial examples (an example that is hard for students of American history to accept), he maintains that the original home of the New Englanders was the hardest environment of all, and that North American history speaks in favor of the proposition that the greater the difficulty, the greater the stimulus to civilization. This thought is in line with his generalization that a new, unmetered physical environment presents a challenge to the affiliated civilization when it has broken new ground, though this principle is one of varying applicability. Even though he discounts environmental theory, the part it does play in the idea of stimulus and response is so crucial that one can without unfairness call it environmentalistic: a real optimum challenge stimulates the challenged party to achieve a single successful response and to acquire momentum from that achievement to engage in fresh struggle, moving from the solution of one problem to the presentation of another.[89]

Historically, the strength and the weakness of environmental theory have been its vagueness, the sloppiness of its formulation, and the arbitrary character of the correlations. It has been assumed that it was something more than a point of view. Stimulating as both the theories of environmental determinism and the possibilist tradition have been to investigations of geographic phenomena (and these propositions cannot be denied for parts of their history), neither can be dignified by a designation more scientific than that they were ways of looking at things, points of approach, of view; a possible means of synthesizing large masses of otherwise recalcitrant material. What could be more gratifying than to have some small bright lamp to guide a man through the complexities, the contradictions, the sheer mass of fact in European history, the idea of showing this history in terms of the geographic factors that had molded it? And though the history would be a different one, what could be more pleasing to other scholars, who could write this history in terms of the environmental opportunities or possibilities open to peoples as they moved into new lands and improved their technology through science and invention? In a very real sense, these approaches, like so many others of an essentially nongeographic character, such as the Marxian approach, or the idea of society developing in response to a theory of progress or cyclical growth, or of history showing advances in ideas of freedom (twentieth-century embarrassments omitted), were sorting devices, each idea having its own special wheat and its own special chaff, to rid oneself of the crushing burden of accumulated facts. The task was easier in a day when it was believed that European civilization represented not only the apex of a long historical development but also a permanent apex. Whatever one may say of this point of view, it had the advantage of eliminating the histories and the national experiences of large parts of the world except as they had contributed in the past to pushing civilization on to its apogee. After they had fulfilled their historical mission, they were allowed politely to rest in the stationary, unchanging inertias to which they were supposedly addicted. The newer appreciation, going beyond the ranks of the specialists and entailing the wider study of the surviving nonliterate peoples of the Earth, as well as the study of non-European cultures, has doomed the simple approaches; it has obliterated the tree markings on the forest paths. This is the essential and honorable failure of environmental theories as over-

---

89. Glacken sprinkles the past two paragraphs with random numbers, ranging from 77 to 89, from Toynbee's *A Study of History*. However, the text does not clearly establish exactly to what assertion any given page number applies. I have therefore omitted specific references.—Ed.

all principles guiding the interpretation of human civilizations and the explanation of individual human cultures. Far more important, both historically and in the present, is what these cultures have done to their environment as a consequence of their settlement history.

In the previous paragraph, I spoke of the passing of environmental theories as overall explanations of human civilizations. These remarks do not apply, however, to certain studies that will continue to be made in the future. Dudley Stamp[90] has commented on this in his remark that the future possibly will be more environmentalist than the past, in the sense that with growing populations, especially in a small country like England, there will be more insistence on the best possible uses of land. We may expect these trends to grow with regional and national planning and conceivably in a world sense, when populations will have increased to such a point that government intervention and planning on a wide and very detailed scale will be part of the new world. The old Hippocratic theories may have died away, but the newer self-conscious environmentalism will be the handmaiden of planned states.

---

90. L. D. Stamp, *The Land of Britain: Its Use and Misuse* (London: Longmans, Green, 1948).

THE MALTHUSIAN SHADOW OVER
THE TWENTIETH CENTURY

This essay also draws on Glacken's handwritten manuscript. It is bookended by two conference proceedings—a 1937 conference on the prospects and limits of land settlement, the proceedings of which were edited by Isaiah Bowman, and the 1949 UN Conference on the Conservation and Utilization of Resources. Sandwiched in the middle are discussions of two books, *The Rape of the Earth*, by G. V. Jacks and R. O. Whyte, and *Road to Survival*, by Glacken's mentor, William Vogt. Thematically, it builds on earlier essays in the book. In introducing the 1937 conference, Bowman framed the problem in terms of the prospects for and obstacles to future colonization. The question then was to explore the limits, both cultural and natural, of what Europeans could do by way of migration. In his summary, Glacken emphasizes Carl Sauer's fascinating essay on the prospect of the redistribution of population. After a detailed empirical survey, Sauer had concluded that "the population of the world has become sedentary permanently, that most of its inhabitants are where they belong."[1]

Next is the discussion of the two books. Jacks and Whyte were two leading British soil scientists who, like many during that period, were preoccupied with the Dust Bowl. Glacken summarizes a book surveying the extant literature on the nature of the habitable world and on scientific explanations of the Dust Bowl before proceeding, like many twentieth-century scientists turned ideologues, to forcefully articulating social and economic ideologies deriving from his interpretation of ecology. Glacken adopts a similar approach to the next book, and here he is remarkably critical of a former mentor, whom he accuses of embracing en-

---

1. Carl O. Sauer, "The Prospect for Redistribution of Population," in *Limits of Land Settlement: A Report on Present-Day Possibilities,* ed. Isaiah Bowman (New York: Council on Foreign Relations, 1937), 7–24. Subsequent quotations in the text are from this edition.

vironmental determinism and the "albatross of Malthusianism." Though the last section of this essay is focused on the UN conference, Glacken directs his attention to one of the twentieth century's most fascinating characters, the British and Australian statistician Colin Clark, who, he argues, was neither a demographic or ecological pessimist nor an optimist but a scholar who drew attention to what Clark described as the problem of rural labor and the phenomenon of outmigration.

The Limits of Land Settlement

The distribution of mankind over the surface of the globe had excited the wonder of man even before the manner of this distribution was fairly well known. The dearth of real knowledge on this subject reached, as we have seen, well past the middle of the nineteenth century. Nowhere is this clearer than in the study of the history of population estimates. To take but one example, two of the better known estimates in the mid-nineteenth century showed reasonably close correspondences to actual population only in the cases of Europe and North and South America. For the rest, the totals differed by hundreds of millions, and completely variant pictures were given of the distribution and density of the world's population.[2] Obviously, the lack of knowledge of the interior populations of Africa and a general paucity of data on Asia and Oceania were responsible.

One fact remained clear, even without a detailed knowledge of the distribution of the world's population: a great deal of the Earth's surface was still very sparsely inhabited. Despite great growth in the world's population, especially since the middle of the eighteenth century, there had been no great change in the relative distribution of mankind on the globe, with the exception, of course, of the New World. How, then, could theories and warnings regarding overpopulation, the possibilities of the Earth filling up, be taken seriously when so little of the world's area was already inhabited? How was it possible, as William Godwin had said of the Malthusian doctrine, to speak of population outstripping the food supply when such a small proportion of the Earth's surface was used for the maintenance of the world's peoples?

One way of answering these questions was to reexamine the nature of the habitable world to discern, if possible, the problems inherent in any future settlement

---

2. *Annalen der Erd-, Völker- und Staatenkunde* [Annals of geography, ethnology and political sciences] (Berlin: V. H Berghaus, 1842); John Downes, ed., *The United States' Almanac, or, Complete ephemeris . . .* (Philadelphia: E. H. Butler, 1842–45).

of the world's "open spaces." This was attempted on a broad scale in a symposium and its resulting book, *Limits of Land Settlement*, edited by Isaiah Bowman and published in 1937.[3] Previous discussions, as we have seen, were concerned with the total available land and the total population it could support. They were concerned only parenthetically with the obstacles such an eventuality would have to overcome. To an even lesser degree were they concerned with the assumptions on which such estimates were based: a certain fluidity or mobility of peoples, a willingness to spread out and colonize new lands, as the pioneers of the nineteenth century had done. They were concerned with results, not with the means by which these results were to be accomplished. Two particularly important phases of the problem were omitted: the nature and history of colonization, and the commonplace observation that, under the influence of modern industry, the whole tendency has been toward the further concentration of peoples, toward higher population densities both in older inhabited lands and in the newer ones, with the concomitant growth of large cities and the general trend toward urban living.

Neither migration nor conquest was an answer to the problem of overcrowded countries. Moreover, while the world's open spaces were open, migration and colonization had to deal with a nationalized world. Colonization was no longer a simple matter, and this was largely a result of modern civilization. "The migrant of 1937 wants civilization to follow him because the homeland is comparatively rich and safe in contrast to the meagerness and limited security of life on the frontier. In earlier classic eras of migration, medical aid was not available by telephone even in the homeland. The greatest losses of the pioneer of those days were friends and familiar scenes," wrote Bowman. This point of view, then, emphasized that the great, unoccupied lands of the Earth are not ready-made environments for man. New environments, it was argued, had to be created with more thorough preparation than had been the case in previous historical settlements. Pioneer development could no longer be haphazard; it had to be the result of planning, of resource appraisal, of adaptation of technologies, of extensions of agricultural knowledge, and of research based on the physiological response to different environments. Further, wrote Bowman, "What these experiments and experiences forecast is that the population capacity of the land depends not on its degree of vacancy, but on the total available resources that land, people, science, technology, and market demand, plus transportation facilities, make possible in combinations that are largely unique from region to region and from country to country."

---

3. Isaiah Bowman, ed., *Limits of Land Settlement: A Report on Present-Day Possibilities* (New York: Council on Foreign Relations, 1937). Subsequent quotations in the text are from this edition.

This statement marked a greater reversal in thinking about the lands still available for human settlement than is at once apparent. Easy generalizations, for instance, based on comparative population densities would have to be discarded. An average density based on some chosen norm could not be applied to the sparsely settled areas whose population was below this density. One could not make simple identifications of "vacancy" and "open spaces" with habitable land. In other words, long-standing vacancy had a historical meaning. There was the implication here too that in the further exhaustion of the habitable Earth, the task was not merely to bring people to new resource opportunities but to develop new lands, bringing new resources to peoples living in the industrialized older settlements. The problem was being oversimplified if the new areas with their new colonists were considered apart from the rest of the world.

The implications of this historically uneven distribution of population are provisionally explored by Carl O. Sauer in the opening contribution to the volume, "The Prospect for Redistribution of Population."[4] He began by observing that "The population problems of the world can be defined only in so far as we have knowledge of growth or decline of human densities as distributed about the world in time." This involved much more than a history of the world's population; Sauer's emphasis was on the comparison of changes by area throughout time, and especially on "far more accurate descriptions of the distribution and 'movement' of population over the world from period to period." This historical and geographic conception is remote from abstract theories of population growth, whether optimistic or pessimistic, and it is equally far from general theories of the man-land ratio, which, as we have seen, had their origin in the early discussions of the nature of rent.

The consequences of dealing with the history of population in these terms can be translated into interpretations regarding the present distribution of mankind. During the period 1492–1918, the most conspicuous fact about population growth was that the world's population had more than doubled since 1800. Sauer wrote, "The period immediately behind us is therefore time that is unique in all history." This means that with minor exceptions, the great resources of the world "have been brought into use" through cultural diffusion. "The expansion of population in the nineteenth century," Sauer said, "is principally a matter of the commercial occupations of the great grasslands of the world." This exploitation of the grasslands, which required "modern transport, heavy horses, horse husbandry, and the modern plow," was made possible by the Industrial Revolution, which in turn owed its full realization to the expansion. The nineteenth-century redistribution of the world's

---

4. Sauer, "The Prospect for Redistribution of Population."

population was thus onto the world's great grasslands of the temperate zones. "The story begins first in South Russia. It is continued in the Argentine, in Western Canada, in Australia, and in South Africa. It ends, and ends for all time, in the Canadian Northwest and the plains of Manchuria and Mongolia."

Sauer believed that this great migration of peoples had, despite some extensions of the habitable area through irrigation, created a vast destructive exploitation in the new lands, and that as a consequence, the subsistence base was shrinking. The remaining areas for settlement in the habitable world were few: in South America, inner Asia, the "Arctic fringe," the tropical forests of the Old and New Worlds. The lack of wide-scale settlement in South America was the result of a solid cause: the existence of the latifundias. Inner Asia, however, was being peopled from both the Russian side and the Chinese side. The remaining areas were difficult to settle. Sauer's conclusion was that the world was pretty well populated, in relation to the present productive skill of man, because the density of population came more and more to approximate known productive capacity. Future colonization, which was most likely in the low-latitude lands, would "depend on a technologic advance, not on old fashioned pioneering." In a sense, then, the world was already filled up, and its population was rather permanently distributed. Sauer thus concluded that "the population of the world has become sedentary permanently, that most of its inhabitants are where they belong."

The distinctions made in this symposium are largely these: that any filling up of new lands of the Earth would come about, under present circumstances (1937), as a result of planned colonization, with close attention being given to psychological, cultural, and physiological factors. This is quite apparent in the contribution of Carl S. Alsberg to the symposium, which set forth the broad distinction between past and present colonization. He argued that historically, "the empty spaces of the earth were peopled gradually over a long period of time."[5] They were peopled largely by subsistence farmers prepared to forgo most of the cultural advantages of the homeland in the hope of ultimate land ownership. Migrants of this type, he argued, were becoming fewer in the world. "The countries that are complaining most of population pressure enjoy for the most part such a level of living that their potential emigrants are rarely willing to accept the lonely, isolated hard life of the eighteenth and nineteenth century pioneer.... In this hard life a monotonous dust is not the least of the hardships." From this, Alsberg was led to an appraisal of the nature of the relatively thinly populated regions of the modern world. After discussing estimates based on contemporary studies, Alsberg concluded that "solely

---

5. Carl S. Alsberg, "The Food Supply in the Migration Process," in Bowman, ed., *Limits of Land Settlement*, 25–56. Subsequent quotations in the text are from this edition.

on account of the cold, approximately a third of the land in the world neither is nor can be used for food production except in a very limited and in fact insignificant scale." A similar situation existed with the world's deserts: "the fraction of them that is now or ever could be irrigated is really very small, even though they total some millions of acres." There remained about 30 million square miles of highly variable quality of land, suitable for agriculture. The semiarid land of the world could not be expected to support dense populations: in fact, more of it may have been put under agriculture than was warranted, and overgrazing and destruction of vegetation might force retreats here. Great, thinly populated areas exist in the tropical rainforest: in New Guinea, Africa, Brazil, and Central America. "It is quite impossible to say how far the rainy forests can be developed in the direction of food production. Surely the process has to be slow, especially in comparison with what is possible on the remaining grasslands of the world."

Alsberg divided the 30 million square miles of land into "the semi-arid, the rainy hot forests, the wet-and-dry hot country, and the temperate agricultural areas." It was not possible, he added, to estimate the amount of land in each, or indeed to estimate the amount of arable land. Even the term arable "is a relative one. . . . Where climate does not absolutely forbid, whether a given piece of land is brought under the plow is really an economic not an agronomic question." In Alsberg's view, the world was in no danger of running short of arable land, and for the population of the Earth as of 1937 there were about 2.1 acres of arable land available per person. He thus argued that there was "for the present, room enough in the world; and a redistribution of population is conceivable such that each family might have enough land to feed itself." But granting that this redistribution could take place, Alsberg doubted its desirability, for it would involve converting "all of the peoples of the world into self-sufficient farmers." A worldwide back-to-the-land movement would involve such capital outlays, he argued, with the measures suggested for alleviating the Great Depression in mind, that they would be "locked up," irrecoverable, and possibly lost. The view expressed here, then, was that there were large areas capable of further settlement, but that the cost of modern settlement was too great to warrant the outlay, and the new lands would fill up slowly as such expenditures, gradually made, allowed.

This view of the habitable world was advanced by men who wrote against a background of World War I, the Great Depression, the insistence of Germany, Italy, and Japan on an outlet for their populations, and evidence of the worldwide misuse of land and natural resources. Alsberg's appraisal of the habitable world is far more sophisticated in outlook than any previously advanced. Older, more naïve ideas of the population increasing until the world was full were submerged in a swelter of new considerations. The thinly populated areas of the world could not be looked on as reservoirs to catch the floods of the overpopulated areas. These lands would have to be completely refashioned and a man-made environment created, bearing in mind the lessons of land abuse. It announced definitely, especially in the words of Sauer, the idea of closed space and the uniqueness of human experience during the past century. Further occupation of the globe would approximate the techniques of advanced cultures: planning and resource appraisal, expense appraisal, the social and political outlook, and the halts called by persisting nationalisms. There was no determinism here, and much modesty. Since little still was known of the amount of arable land in the world—or even what it was—the reasons why some tropical areas had remained unpeopled were unknown. Guesses and trends were the chief tools at hand. It is a view that came from stressing the history of population distribution (not relics of population growth), from an appreciation that environments can change for the worse as well as for the better, and from the extraordinary difficulties interposed by human culture, nationalism, and economic considerations. In this sense, it was neither optimistic nor pessimistic. But the simplistic views relating population, food supply, and the Earth filling up like water overflowing (drowned valleys) into valleys linked together by passes in the mountains were gone. But they were to be revived in a more sophisticated way in the discussions that followed World War II.

Soils and Civilizations

Although an immense amount of material had accumulated on the subject of soil erosion and conservation prior to the 1930s, it was during the latter part of the decade that the subject again attracted worldwide interest. I say again, for Marsh's work had also attracted considerable interest, as had the dramatic reports of President Theodore Roosevelt's committee on conservation.[6] Even for those now living who were adults during this period, a rereading of a sampling of the literature is a reminder of the dramatic character of much of this writing. It was probably the great Dust Bowl storms of the 1930s that touched off the new realization. Most of the comment centered on soil erosion. In the United States, Paul Sears's *Deserts on the March* appeared in 1935.[7] The work of Dr. H. H. Bennett and Walter Lowdermilk on soils and agricultural practice had further dramatized the "new" problem. The reports in America of the National Resources Board, the work of Carter Goodrich and his associates on migration, the publication of the Department of Agriculture's 1938 Year Book, *Soils and Men,* and the writings of Charles Kellogg had, before

---

6. Glacken is likely referring to the seven conservation commissions and conferences established by Roosevelt during his presidency, 1901–9.—Ed.

7. Paul Sears, *Deserts on the March* (Norman: University of Oklahoma Press, 1935).

the American entry into the war, given strong professional sanction to the movement to save the world's soils.[8] New themes appeared, such as Carl Sauer's idea of a rhythm to man's harmony and disharmony with nature.[9]

It is probable that the occurrence of these dust storms in America during a period of deep, worldwide depression and the great internal migrations—with the attendant literature of protest, as in the world of Steinbeck, which was a response to both conditions—were responsible for the deep and lasting impression they made on America and the world, for American leadership in soil conservation matters was soon established, and the foreign literature is full of references to this experience. But I do not wish to trace this history but to point out that a work on soil erosion and the disharmony of man with nature, the relation of soils to civilization, was certain to appear as a result of these disclosures. Not only did attention turn to the United States, but there was a greater stimulus to look at local and scattered erosion as part of a worldwide pattern. Such a synthesis appeared in 1939. This was the work of two British soil scientists, G. V. Jacks and R. O. Whyte, authors of an important book on soil erosion globally, *The Rape of the Earth*.[10] There are many different themes in this work, some old, some new, as they arose out of contemporary conditions. Many of them were to be repeated by subsequent writers. Too often works of this type are cited more for their technical information on the geographic conditions they describe. The ideas and the interpretations, scattered throughout, are neglected, as if they were personal matters. Since this work is extraordinarily rich in its collection of ideas, like Vogt's book later, it is this portion of their work that will be discussed, for it contains a description of the nature of the habitable world today, explanations of the disappearance of areas now deserted but formerly habitable, and a series of extraordinary ideas concerning the relationship of economic nationalism to soil conservation.

In the first place, the rise and fall of civilization is flatly explained by Jacks and Whyte as a result of soil erosion. World history is interpreted as the history of soil erosion. "One after another, the great empires and civilizations of the past have been swept away by soil erosion." This is entirely different from earlier themes of

---

8. U.S. Department of Agriculture, *Soils and Men, Yearbook of Agriculture* (Washington, DC, 1938). [Dr. Charles E. Kellogg was the third chief of the USDA's Bureau of Chemistry and Soils and helped shape the National Cooperative Soil Survey Program. He wrote, among other things, a classic, *The Soils That Support Us: An Introduction to the Study of Soils and Their Use by Men* (New York: Macmillan, 1941).—Ed.]

9. For a good review of Carl Sauer's ideas, see William M. Denevan and Kent Mathewson, eds., *Carl Sauer on Culture and Landscape: Readings and Commentaries* (Baton Rouge: Louisiana State University Press, 2009).—Ed.

10. G .V. Jacks and R. O. Whyte, *The Rape of the Earth: A World Survey of Soil Erosion* (London: Faber and Faber, 1939). Subsequent quotations in the text are from this edition.

soil exhaustion as an influence on history. Here the process involved is one of expansion of empire, followed by *soil exhaustion* as a result of excessive demands put on soils, and this *soil exhaustion* prepared the way for and was followed by *soil erosion*. Erosion is thus a far more advanced phenomenon than soil exhaustion, and is more permanent and irreparable. Earlier writers had thought of exhaustion in terms of less of the essential minerals or organic materials, but with none of the irreversibilities associated with erosion. Northwestern China, Persia, Mesopotamia, and northern Africa are cited by Jacks and Whyte as early historical examples of erosion following exhaustion. "The decline of the Roman Empire is a story of deforestation, soil exhaustion, and erosion. Judea, likewise, was overcome by erosion." This universal explanation they apply to the old civilizations of Asia and the New World. In Central America, Colombia, and Ceylon, "those civilizations were exterminated completely when man could no longer hold [his] own on soil." Not one shred of evidence is offered to substantiate these assertions, which, if true, would at long last provide the proper key to the interpretation of history. They assert: "The shot fired at Lexington that was heard around the world in 1775 had less influence on subsequent world history than the red dust from Nebraska that settled on Washington in 1934 will have on the future." Here one could make a good case from the authors' own work that the 1775 shot had a tremendous influence on the 1934 dust storms, which would give it equal historical rank, to say the least.

In modern times, this interpretation is carried forward, it must be admitted, with considerably more evidence than the broader historical generalizations were. Since so much of modern history is concerned with European expansion and influence, the experience of peoples of Eurasian origin in other parts of the world is of more than regional interest. This accounts for the authors' emphasis on the unique susceptibility of European-type civilizations to induce widespread erosion on non-European lands: "Man-induced soil erosion is taking place today in almost every country inhabited by civilized man, except Northwest Europe. It is a disease to which any civilization founded on the European model seems liable when it attempts to grow outside Europe." The chief distinction made here is that through long experience, Europe had developed an intensive agriculture and a system of cultivation that has "*under European* conditions increased soil fertility," and that "Europe owes its immunity from soil erosion to the adaptation of its agriculture to its climate;" while in the New World, the conditions were entirely different. Here, virgin soils were exploited for immediate profits induced by the progress of agronomy, farm machinery, and "the insatiable demands of the Old World." Later, these authors, as East had argued as well, point out that European soils had been subject to no great strain in the whole history of the world until World War I: "The enormous increase in Europe's population was accomplished without disturbing the established equilibrium between humanity and the soil, although . . . it has clearly prevented equilibrium from being established elsewhere. . . . Whether for her ultimate good or not, Europe will have to rely to a far greater extent than hitherto on her own soils to feed her teeming populations. For the first time since they came under cultivation, the soils will be subjected to a serious strain." This statement, a reasonable prediction in the late thirties, has of course been invalidated by the continuing dependence on non-European soils since the war.

This theme is further developed. Jacks and Whyte argue that the Old World has been essentially parasitic on the New; that worldwide trade and freedom from economic barriers are a menace to the world's soils; and the more general theme, that erosion is the symptom of maladjustment between human society and the physical environment. This parasitism of Europe, to take up the first point, is regarded in a manner reminiscent of Friedrich's theories:

> The main economic cause of recent accelerated erosion has been the transfer of capital across regional or political boundaries and its repayment with soil fertility. Had Europe solved its population problem by exporting more men and less capital and therefore neither needing nor being in a position to buy so much food from abroad, the world would have been poorer in many ways, but there would have been more soil left to posterity. Progress would have been slow and there might have been no erosion problem to tax man's ingenuity and courage to the utmost and spur him to greater conquests.

This is interesting both for its assumption that the export of Europeans would have resulted in a smaller population in Europe and for the implication that mankind will not progress without problems of the magnitude of soil erosion to grapple with. Friedrich, it will be remembered, had argued that destruction, bad as it was, was necessary, for it was this devastation that spurred men on to greater achievement.[11] The relationship of economic nationalism, world trade, and soil erosion, as described in this work, is probably one of the starkest juxtapositions yet made between ideas of economic prosperity based on lack of trade barriers and ideas of soil conservation based on maintaining the habitability of the world. The threat of war, existing nationalism, and autarchy are seen as allies in the preservation of the world's soils.

It has usually been assumed that population pressure, when it is considered, only serves to accelerate soil erosion and the dissipation of natural resources. This position is typical of many writers after World War II but was by no means the case prior to the war. It was often considered, on the contrary, that population pressure

---

11. For more on Friedrich, see the fifth essay in this book, "Key Trends in Nineteenth-Century Environmentalism."

forced a people to take necessary conservation steps. "It is unlikely," write Jacks and Whyte, "that any effective and lasting organization to protect the soil from misuse and erosion will evolve in a country where the pressure of population on the land is small or non-existent." These authors feel that the peoples of the Earth meet the emergencies caused by erosion only when faced with a growing population within their own borders, or with shrinking arable land and expanding desert land. Speaking of Canada and Australia, Jacks and Whyte write: "Naturally it is hoped that in future the stimulus to undertake the process of radical readjustment, which accelerating erosion shows to be inevitable, will come from the pressure of expanding populations on the land, and not of expanding deserts on the people, but both countries appear to have a long way to go yet before the stimulus becomes powerful enough to be translated into action."

In the twentieth century at least, writers who discuss the relationship of soils and conservation to civilization are soon brought to two further subjects: the role of science, if such a general term can be used, and the effect of the urbanization of many peoples of the world—another product of science and the machine age. Though no generalization is possible, it is noteworthy that quite often the most cautious estimates of the power of science come from scientists themselves, though this caution may be more pronounced among biologists, soil scientists, and ecologists. Jacks and Whyte, for example, take the view that there are limits to applied science in this field, to say nothing of the fact that much scientific progress has occurred along lines that indirectly have worsened the soil erosion problem: "There is a limit to the extent to which applied science can temporarily force up soil productivity, but there is no limit except zero to the extent to which erosion can permanently reduce it." One of the important historical events of modern times, they felt, has been the denudation of the semiarid continental grasslands of the New World and the erosion consequent on this loss of a grass cover: "It is a humbling thought that modern men, with their immense knowledge and capabilities, have been defeated by grass, whereas their ancestors succeeded in the more difficult task of subduing the forest. But they took a thousand years or more over the job; their modern descendants have acted as though, with their machines, they could conquer the grasslands in a century." Machines, in this conception, are regarded as an expensive and false substitute for cultivation of soil and as producing a dangerous psychological distance between the urban man and the soil.

> The respite which, with the advent of the machine age, a large part of the civilized world gained from its arduous primary duty of cultivating the soil has been purchased at great cost. Machines offered to them new lives, at once more lucrative and physically less tiring than agriculture, and started the drift from country to towns that has con-

tinued with accelerating momentum to the present time. As the countryside became depleted, machines in the towns intensified the demands made on the land; machines on the railways, roads and sea enabled the increased demands to be satisfied without difficulty; machines on the land took the place of man as cultivators. This accelerating mass production from the land has loosened, and in many cases severed, the biological link between the cultivator and the soil.... The townsman takes at least a dilettante's interest in the land; the delicate balance of nature upon which the stability of civilization depends can have no meaning to him because he has completely conquered and divorced himself from Nature within the ordinary orbit of existence. This becomes a serious matter when more than half the civilized world lives in towns. One result has been soil erosion insidiously creeping over the world, and finally and inevitably forcing itself upon the townsman's attention.

This idea of town life as walling off mankind from nature was forcibly stated by Woeikof in 1901. Jacks and Whyte have added to it an additional idea, that the self-incarceration of urban life has been so effective that worldwide erosion progressed to dangerous proportions before urban man was even aware of it.[12] Nature, in this interpretation, is not thought of in the Malthusian sense. It is not niggardly, nor does it limit human growth. But it would be a mistake to think of civilization as superposed on the Earth, like a house on a lot. "Probably more soil was lost from the world between 1914 and 1934 than in the whole of previous human history. By 1935 the illusion that nations could get rich quick at the expense of a beneficent, unresisting Nature had formally been shattered."

In their survey of the geography of soil erosion, Jacks and Whyte make a distinction between the eroded new lands of the world, such as in America and South Africa, and lands where erosion has been the result of long occupancy, such as in northern China, arguing that "prolonged rather than excessive utilization has ruined the land round the headwaters of the Yellow River in China—the worst eroded regions in the world." To preserve the African continent from soil erosion, the authors envisage a new feudalism, under white leadership:

> A feudal type of society in which native cultivators would to some extent be tied to the lands of their European overlords seems the most generally suited to meet the needs of the soil in the present state of African development. South Africa would become less attractive with greater dependence on its soil. The urban white population cannot hope to cultivate successfully these exhausted lands. The white people should increase in power, but decrease in numbers; the opposite fate awaits the black people. Only

---

12. Glacken was likely referring to Alexander Ivanovich Woeikof's *Sustainable Development of Rural Systems* (1901).—Ed.

thus can Nature's harsh harmony, so rudely shattered by the advent of civilization, be restored.[13]

The development of tropical lands involved contradictions of a far-reaching nature. The fertility of these regions lies in the vegetation itself, not the soils. Use of the soil must therefore be of a temporary nature. "It must be admitted that no agricultural system except shifting cultivation has yet been devised that will ensure lasting stability and fertility to tropical forest soils under human management."

In this view, virtually no part of the habitable world has escaped devastation so thorough that civilization itself is threatened. Long occupation and slow exhaustion in the old countries, short occupation and quick exploitation in the new countries, the peculiar requirements of tropical cultivation, the delicate balance in Africa, and the conflicting claims of white and native culture are leading to a more confined world and forcibly, through nationalism, perhaps even to a new feudalism, to a worldwide attempt to adjust to the conditions umpired by the environment. This environment is considered in terms of soils, vegetation, and climate. A realization of the nature of this environment is creating a new era in world civilization. The soil conservation movement, Jacks and Whyte argue, "steadily gathering momentum throughout the New World, represents an early stage in civilization's adaptation to new environments." They argue that this new direction in social evolution cannot come about through dictatorial legislation; it will probably come about through common consent, based on an appreciation of the urgencies involved. "One can envisage the slow building up of a complex society based upon the national interdependence of town and country, and the hills and valleys and plains that give rise to a great river and determine its few."

This was a synthesis that appeared on the eve of World War II. It is fresh with the impressions of the American Dust Bowl, with the trade barriers and the emphasis on national self-sufficiency that were so prominent in the thirties. It is fresh with the TVA and the beginnings of a world conservation movement. In some ways the problems discussed are no different today: the continuing emphasis on the contradictions between human culture and soil conservation, the destruction of ecological balances. But it is a doubtful whether the authors, or anyone else, would now see in intense economic nationalism the main avenue to preserving the habitability of the world. It is true that war and the threat of war, and an even more intense economic nationalism, might enforce rigid conservation measures in all countries regardless of political structure, as a means of national survival. But it is also true that these

---

13. For a discussion of such ideas with the Apartheid regime in South Africa, see Peder Anker, *Imperial Ecology: Environmental Order in the British Empire, 1895–1945* (Cambridge, MA: Harvard University Press, 2002).—Ed.

conditions, as happened in Germany and the German-occupied countries, might produce the opposite effect in a frantic attempt to reach a decision. In any event, this work, like others before it, and others that followed and will follow it, clearly shows how questions concerning the nature of the physical environment and man's control over it have been woven into the political, economic, and social conditions of our time. Gone—it would seem forever—is that simple mainstream of thought that saw in nature the determinant of history. Civilization is now regarded in terms of abilities to preserve its physical environment, not on a local but on a world scale.

The Road to Survival

One of the most important ideas to emerge, in full origin, in the aftermath of World War II was that mankind the world over was engaged in destroying the environment and in making the world uninhabitable. None of the ideas advanced was new, but the combinations of ideas advanced were. The trenchant analyses that came out of World War II were based on three sets of observations: (1) there was insufficient food in the world, (2) the world's population was increasing, and (3) the combination of these two circumstances was accelerating soil erosion and the wasteful use of resources on a worldwide scale. This literature in general differed considerably from that which followed World War I, in which the emphasis was almost entirely on the increase in population along Malthusian lines. The problem then was seen largely in terms of overpopulation and the inability of the world's agricultural lands, even with admitted possibilities of technical improvement, to sustain the world's peoples. It has been said that the Great Depression alleviated the fear of overpopulation and, for a time, even gave place to a depopulation scare. This may be true in the sense that it was widely believed that it was the lack of cash rather than the lack of food that imperiled (at least) the Western world. Fear of overpopulation, if it did exist, was probably more of an occupational fear, or the fear that there were too many people or too few jobs, from day labor to professional jobs, and that overpopulation could be discerned in certain age groups as a result of the limited opportunities for the middle-aged and older segments of the population. If food was not lacking—though available only through public distribution—at least worldwide want was present. The most extensive worldwide investigations into the extent of erosion and deterioration of the environment were undertaken in the thirties. Much of this work was the result of the American Dust Bowl experience, but the widespread worldwide interest in conservation was probably unmatched in any previous period, with the possible exception of the investigations undertaken by the Roosevelt administration of the American situation at the beginning of the twentieth century. Later investigations of the period have shown, however, that large areas of the world had insufficient food even before the war.

Much of the literature since the end of World War II, totally aside from the question of peace and war, has been pessimistic regarding the ability of the human race to maintain a habitable world, in view of the accumulation of social and political problems, some of almost prehistoric origin, and of the growth of the world's population and its implications for the preservation and rational use of the world's resources. One of the sharpest and most pessimistic of these works was William Vogt's *Road to Survival* (1948).[14] There were, of course, many others that advanced similar—though not identical—themes, for example Ezra Parmalee Prentice's *Hunger and History* (1951), Henry Fairfield Osborn's *Our Plundered Planet* (1948), and Frank A. Pearson and Floyd A. Harper's *The World's Hunger* (1945).[15] I have chosen Vogt's book here not because it is considered representative of postwar thinking on the subject but because it has drawn on a greater variety of difficult ideas (each with its own history) than have others with a similar viewpoint. The book has, of course, been widely praised and attacked, and both reactions, in my opinion, owe less to the subject matter than to the manner of presentation. But for our purposes, it does not matter whether the book is optimistic or pessimistic; my intention is to examine the nature of the present and future habitable world that emerges from Vogt's analysis and to trace the ideas out of which this view was constructed. In this dispassionate way, I believe certain strengths and confusion will be apparent, for few would argue the thesis that, even with the prospects of prolonged peace, the problems centering on population and the maintenance of resources will continue to occupy the thought of people everywhere.

Four principal ideas make up Vogt's description of the nature of the habitable world: (1) the catastrophic growth of the world's population, (2) worldwide soil erosion, largely a result of disturbances in the hydrological cycle, (3) the deleterious effect of certain historical events, with an emphasis on cumulative effects since the nineteenth century, including the effects of technological improvement, medical research, customs, and national traditions, and (4) confusions in thought and in the meanings of various words at different levels of abstraction, which may ultimately be traced back to Aristotle. It will be interesting to follow the ideas in the order in which they appear in Vogt's argument. He considered the possibilities of great technical improvements, but felt they could not come in time to correct present disharmonies. In dealing with the relationship of man to his environment, he argued that the assessment should rest not on what can be achieved ideally but

14. William Vogt, *Road to Survival* (New York: William Sloane Associates, 1948).

15. Ezra Parmalee Prentice, *Hunger and History: The Influence of Hunger on Human History* (Caldwell, ID: Caxton Printers, 1951); Henry Fairfield Osborn, *Our Plundered Planet* (London: Faber and Faber, 1948); and Frank A. Pearson and Floyd A. Harper, *The World's Hunger* (Ithaca, NY: Cornell University Press, 1945).

on what can be achieved practically. "For it is the least favorable conditions, not the most favorable," he writes, in a manner suggesting a generalized application of Liebig's law of the minimum, "that determine the ability of the earth to support human beings."[16]

Vogt maintained that there exist what he termed "practical ceilings," as contrasted with theoretical ones, and that these ceilings are imposed by "environmental resistance," which he defined as the sum of the varying limiting factors acting on the biotic potential. This idea was developed in ecology as an outgrowth of Liebig's law of the minimum to account for various growth limits—restrictions on plant and animal populations imposed by the nature of the environment.[17] The following illustration shows one phase of this general idea. In studying an animal population, Vogt said, reproduction has a certain potential maximal value, and when mortality numbers are subtracted from it, the difference represents the actual size of the population at the moment. This mortality component, Vogt argued, is controlled by "environmental resistance" and may be stated as a simple equation: "Population growth = potential reproduction—environmental resistance."

Vogt expanded this idea considerably by the use of a similar formula: $C = B : E$, where C is the carrying capacity of a given area, B is the biotic potential, or the ability of the land to produce plants, and E is environmental resistance. In simple terms, the carrying capacity is the result of the ratio between the other two factors. While Vogt conceded that the equation is oversimplified, it should be pointed out that the use of the ratios demands a high biotic potential to achieve a moderate carrying capacity. In any case, this is a technical statement of the more general idea that population, whether of people or of plants, is seriously limited by environmental factors. In the equation $C = B : E$, Vogt's argument is that the activities of man tend to increase E, and therefore, if B remains constant, C will fall, or that any increase in E will require a much greater increase in B, since the equation is in the form of a ratio. If, for example, the biotic potential of a plot of ground is 500 ears of corn and the environmental resistance is such as to reduce the potential to 250 ears, C has a value of ½. If human interferences such as soil erosion or the encouragement of insect life increase the resistance so that only 100 ears can be produced, then C goes to ¼.

It would seem that this formula adds little to the qualitative descriptions of the relationships that appear in Vogt's work. With this relationship established, it then

---

16. The law of the minimum, named after Justus von Liebig's work, states that agricultural growth is controlled not by the total amount of resources available but by the limiting factor. —Ed.

17. There is a suggestion, in a side note, that Glacken might have written a full chapter on Liebig.—Ed.

becomes important to estimate the amount of arable land available on Earth. As the illustrations so far given have shown, this is a crucial estimate to make, for on the amount of the estimate (aside, of course, from political and social questions) depends much of the optimism or pessimism of such views, especially with reference to a future in which populations presumably will continue to grow. One of the chief mistakes, Vogt felt, was the tendency to expand into areas not suited to agriculture. For this reason, Vogt argued that malaria was a blessing in disguise, since a large proportion of the malaria belt was not suited to agriculture.

These points of view, in the main, rest on the general ecological idea of the equilibrium in a balance in nature, and that interferences, especially of a drastic sort, are harmful to the equilibrium and will result in the destruction of the environment as far as its human habitability is concerned. This idea is at the base of his sharpest asperities, a point often forgotten by admiring and enraged critics alike. This equilibrium, Vogt claimed, is characteristic of all nature, even with primitive man in the picture. He argued further that the combination of intelligence and knowledge and population growth had accelerated the disturbances of the equilibrium in a drastic way, compared to the scarcely perceptible ones of prehistoric times. Population growth was particularly bad, he held, because it increased destructive exploitation and the use of lands unsuited for cultivation.

Vogt argued that the great population increases of the eighteenth through the twentieth centuries created "a vast school of new limiting factors," largely through a sharply accelerating command of nature. He credited Malthus, Franklin, and Jefferson with a partial comprehension of what was happening but claimed that not even Malthus foresaw that in the core of increasing "production" was hidden what Vogt called the "the worm that would finally consume the earth." I hope it is not destroying Vogt's views to say that we have two worlds, a primeval world in equilibrium and a world of economic activity, and that modern history especially is the history of increasing displacement of the fruit of the second, through the instrumentality mainly of population increase, to the point where the first will be destroyed, at least for the purposes of complex civilizations. The view of history expounded here is thus a combination of the idea of harmony in nature, as old as the Greeks, enormously enriched by the biological and ecological thinking of the last hundred years, and the idea of man as a destructive geographic agent, in the tradition of Marsh, Shaler, and Woeikof, with a Malthusian interpretation of population growth.

Vogt argued that the activities of European man during the nineteenth and twentieth centuries were chiefly responsible for initiating the worldwide destruction of the environment. He argued that primitive man, and even some highly cultural groups of men, had historically destroyed their environment. However, they had never done so "with the seemingly calculated inexorability of a Panzer division." The steel ax, the moldboard plow, fire, and firearms, he argued, were the chief tools in this destruction. A little further on, Vogt brought up the subject of anthropocentrism, claiming that "so anthropocentric has he [man] been that, since he began to achieve what we call civilization, he has assumed that he lives in a sort of vacuum. He has probably been an agriculturist at least a hundred centuries, yet it was only in 1944 that Dr. E. H. Graham published his first book on national principles of land use."[18] The history of ideas does not bear out this interesting analysis. For instance, the relationship of man to his physical environment has been the subject of speculation probably as long as discussions of the nature of truth have existed. For more than two thousand years there has been a continuous literature on the subject. The trouble is not that man "has assumed that he lives in a sort of vacuum" but that the assumption has been made that man is at the mercy of his physical environment; that its climates, the contours of its continents, its temperature or humidity have controlled human destinies. The literature begins with Hippocrates and Aristotle, Thucydides, Herodotus, Pliny, and Golen and continues in the work of Bodin, Montesquieu, Herder, Ritter, Ratzel, Griffith Taylor.. Environmental determinism in fact has been so strong that only with the greatest difficulty have systemic studies of man's role in changing the physical environment been made. For historical reasons also, the type of ideas on which Graham's book was based are achievements of the nineteenth and twentieth centuries, especially knowledge of soils and ecological principles, a full knowledge of which had to await a real exploration of the world, which was not possible until the nineteenth-century transportation system made it possible. Even in ecology, which lies at the base of Vogt's interpretation, the idea of the struggle for existence was so strong that there was little danger that weak man could destroy strong nature. The real dangers were seen more clearly by the geologists. Probably on the intellectual and scholarly level, the greatest single barrier to an understanding of the relationship was the intense power which the Malthusian doctrine, through the Darwinian theory of the struggle for existence, gave to nature. In a world like this, there was little room, except among the few persons divorced from the mainstream of thinking, to see how man could be destroying nature.

Vogt further illustrated his thesis by selecting the doctor, nationalism, and Aristotelian logic as all contributing to the same ultimate end of destroying the equilibrium of nature. The doctor, by prolonging life and cutting down on the death rate; nationalism, by artificially inducing further exploitation of a nation's resources; and

---

18. Vogt was referring here to E. H. Graham, *Natural Principles of Land Use* (Oxford: Oxford University Press, 1944).—Ed.

Aristotelian logic, by failing to see relationships in their context, such as the ecological point of view, Vogt claimed, all contributed to the failure to conserve the environment. Vogt's book, unlike any other that had been written before on this subject, insists on the importance of the "science and sanity" of Alfred Korzybski.[19] This is too involved a subject for this discussion, but the point is that the world's plight can be traced to semantic confusions. Vogt used as an example the word "land": it may be thought of at the submicroscopic level as "colloidal," atomic, etc.: at the microscopic level its structure and the kinds of microorganisms it contains can be determined; and at a higher level, the macroscopic, there is further abstraction, but it is still the level of reality, for, [it is] at this level that we maintain soil surveys, agricultural experiment stations, agricultural colleges, and so forth. Vogt claimed that these three levels "of reality" are limited to things that someday, if not yet, may be weighed, or measured, in fact, or tasted, or seen. But they may not be *said*. They exist at what Korzybski called the "inoperable level": when we begin to talk about them, we necessarily use a higher-order abstraction. At this first verbal level of abstraction, Vogt argued, land is conceived of as real estate or as a house lot. He claimed that it is difficult to differentiate between the *word* land and the *process* in the *object*, land, and that the two are identified in thought and speech. He wrote that this partially explained why thousands of unfortunate refugees were being dumped into tropical countries that did not have "nonverbal land capable of feeding their own people."

There is little sense in quarreling with those who wish greater precision in the use of words, especially when so many single words may have different meanings. But I wonder if this is a problem that can be solved by the Korzybski and Vogt approach. There are many such words, such as freedom, liberty, romanticism, and nature. Much of this confusion can only be explained and made clear by reverting to the history of ideas. Most meanings of words, especially those that carry an emotional content for the age in which they are employed, have precise historical reasons for acquiring their significance. I have in this study called attention to the varying ideas of nature, to say nothing of the idea of land as used by sociologists and economists and the changing concepts of soil. One of the most confused of these terms is one, by the way, that Vogt used as if it had a meaning assumed by all: "the laws of nature." Yet it would require a sizable volume to treat the nuances of this phrase adequately. The same thing applies to "environment." Much of this difficulty has arisen historically, and will probably continue. For one thing, disciplines create their own vocabularies, and these become traditional; definitions differ according to profession

---

19. Alfred Habdank Skarbek Korzybski (1879–1950) was a Polish American scholar of semantics.—Ed.

and trade. It has been said, for example, that an engineer could be writing about clays and a soil scientist about silt, but they could both be talking about the same substance, because the size in microns for the two varied in each definition. Even at best, the approach is limited geographically, for the non-Aristotelian countries such as China and India, for example, have essentially the same problems.

The Korzybski approach, for example, did not save Vogt from unnecessarily hanging the albatross of Malthusianism around his neck. While he nowhere committed himself to Malthusianism, it is clear from the book that he regarded Malthus as an enlightened predecessor in perceiving at least one phase of the relationship of man to his environment. Now, Vogt could have immeasurably increased the force of his argument and avoided confusion by having his effects catalogued as neo-Malthusianism if he had attacked the whole Malthusian doctrine as leading the learned world into an abstraction that had little relationship to Vogt's central problem, that of preserving and maintaining the world's habitability. Malthus, it is true, saw all the evils of society as caused by population outstripping food supply, but fundamentally, for him, such a trajectory was necessary and desirable. It had to be this way; it was a natural law; it was the way Earth was peopled; it was the way mankind wrested a living in the struggle for existence—Malthus used this precise phrase—with niggardly nature. This was the reason for his lifelong opposition to contraceptive devices and artificial birth control. He had no choice. To have accepted artificial birth control would have removed the underpinnings of the whole theory: the fundamental necessity, despite all the misery it gave rise to, of the tendency to increase beyond the means of subsistence. Malthus had moral grounds for opposing this as well. This is the reason why it is so ludicrous to call those whose main weapon against population is artificial birth control neo-Malthusians. They might as well be called Shakespearians, Habermaseans, or Neoplatonists as far as the logic was concerned.

In the Malthusian doctrine, the environment appears as a challenge to mankind. Man's progress is measured in terms of his ability, through this beneficent yet devastating power of increase, to obtain a living from a stingy nature. It is time that continence and late marriage should be advisable, but not to the extent of abolishing the only real urge that presses mankind on in its fitful progress, for Malthus was no believer in the inevitable perfectibility of man. This doctrine has little of value to a twentieth-century scholar whose theme is that there must be an understanding between man and nature, not a struggle with it. Vogt's argument would have been stronger and less confused had he presented his case as a problem arising out of the observed and irrefutable increase in the world's population since 1750 (there is no Malthusianism here), and the equally demonstrable urgency of a worldwide conservation program.

The chapter titled "Industrial Man—The Great Illusion" contains the entire interpretation of history on which *Road to Survival* is constructed. While Vogt said elsewhere that we continue to live by ideas evoked twenty centuries ago, the actual historical events that led to the destruction of the environment began in 1492, with the most important events being the agricultural and Industrial revolutions. The theme, a familiar one by now, was that the soils of the New World, the agrarian society in America—Vogt seemed to view this history largely in terms of Europe, especially England, and America—furnished the wherewithal for the Industrial Revolution to continue, since the amount of land in Europe did not increase. Vogt argued that speculators and entrepreneurs made the wheels spin faster, but that "it was the rich forests of New England, the prairie soils of Illinois, the red lands of Georgia and the Carolinas, the slopes of São Paulo, the black soils of the pampas that kept them from slowly grinding to a stop."

It was the discoveries of Pasteur and the sanitary revolution that followed that caused the great population increases. With this, Vogt arrived at one of his chief contentions, that industrialization is primarily parasitic on the land. This, as he recognized, was a return to the philosophy of the physiocrats. When British industrial society of the nineteenth century is considered as a parasite on the world's lands, or when, as in the long list of occupations he gives, all pursuits are considered as based in the produce of the land, the conception is little different from the statements of the physiocrats at the end of the eighteenth century, that civilization depended on the ability of agriculture to create a net product. Practically all of Vogt's arguments regarding industrialization are a twentieth-century restatement of this fundamental physiocratic idea that land is the sole source of all creativeness and that all other things, useful as they may be, are sterile compared with it, in fact parasitic on it. Vogt, of course, carried the argument further: that the land itself was being ruined and exhausted to continue feeding the parasite of industrialization. "Had the parasite of European industrial development not been able to sink its proboscis deep into new lands, world history would have been very different. Enormous populations, heavy industry, social and economic pressures could not possibly have developed into the great carbuncle that exploded as World War I."

Vogt attacked what he called the fallacy of industrialization from three different ideas: (1) the physiocratic idea that agriculture is the only creative industry, (2) the idea that economically, industrialization, when food supply became a matter of imports, did not result in an improved standard of living, especially when a large population increase had accomplished the process, and (3) the idea that industrialization had removed man from an awareness of his place in the environment. He admitted that industrialization had made "it possible during a hundred years for the most powerful sector of the human race to live as though it was independent of the earth." However, a high standard of living is not possible for the world's peoples: "By the use of the machine, by exploitation of the world's resources in a purely extractive basis, we have postponed the meeting at the ecological judgment seat. The handwriting on the wall of five continents now tells us that the Day of Judgment is at hand."

Modern man, especially urban man, Vogt felt, is unaware of his relationship to rainfall, and this is most dangerously apparent in "the shattering of the hydrologic cycle," first in the destruction of the plant cover, then in the washing away or blowing away of the topsoil. He writes that man "is the only organism known that lives by destroying the environment indispensible to his survival" and that "the villain in this human tragedy is the primarily uncontrolled rain-drop—which is why I maintain that it is probably the most important single factor influencing man." Another disastrous break in the hydrological cycle Vogt found in modern sanitation: "One of the greatest obstacles to the survival of modern man is his highly developed system of sanitation, which every year sends millions of tons of mineral wealth and organic matter, taken from his farms, forests and grasslands to be lost in the sea."

Vogt completed his analysis with a survey of conditions on the five continents. It is sufficient here to say that for varying reasons or for different combinations of reasons, he found the problems of population, interferences with the hydrological cycle, and ecological imbalance worldwide in scope. Persons interested in his sources and analyses must turn to the book itself, for the material cannot be easily compressed. An important corollary to this world survey—and one that adds to the pessimism of his view—is that the thinly settled areas of the world, the tropics, Africa, and so on are poorly situated for supporting large populations, and that the hope of using them as a real solution is an illusion. According to this view, the environment is shrinking rather than expanding. His own research in Central and South America, done for the Pan American Union, and the work of Harvey in Africa are the chief sources of his doubts as to the hopefulness of new areas.

In conclusion, no man simply describes what he sees or sets down indiscriminately what he has read about a single area or the entire habitable world. To do either would not be possible. Vogt's view, like all others, therefore is a view relying on a selection of ideas. Some of these are relatively new, some are quite old. His view is made up of building blocks from many sources: a view toward agriculture and industry that is physiocratic, a view toward population that is sympathetic to the Malthusian doctrine, an interpretation of the nature of tropical soils that precludes the use of these regions to solve problems of human expansion, an interpretation of modern history as on the whole a disastrous influence on ecological relationships, and the scientific confusions in the meaning of words that modern complex society has produced, together with an insistence that human culture, especially in its urban concentrations, has erected a world, to its peril, independent of the physical environment.

From reading Vogt's book, one would think at first that this is a new form of environmental determinism. The first chapter is titled "The Earth Answers Back." Actually, it is not. Older forms of environmentalism that one meets up with so often in eighteenth-, nineteenth-, and twentieth-century writings saw in mankind the expression of environmental influences, whether they were climatic, geographic, locational, or owing to different soils. In all these works, the supremacy of nature, nature's superior ability to mold human culture, is never questioned. In this sense, Malthusianism also is an environmental theory. But Vogt's is not. His whole theme is that man is a geographic agent of great power, and here, as I have said before, he is in the tradition of Marsh, Shaler, and Woeikof. The difference can be seen clearly when we recall the work of Julien-Joseph Virey. When human interferences went too far, nature lashed back and restored the harmony. In Vogt's view, the environment is being literally destroyed. Like all interpretations of history, this view will be accepted or rejected largely to the degree that the individual accepts—or is persuaded to accept—the ideas, chosen out of the welter of possibilities, that are combined to form to the composite view.

The UN Conference on the Conservation and Utilization of Resources

More than two hundred years separate the world appraisals of the abbé de Saint-Pierre, the author of a plan for perpetual peace, and the UN Conference on the Conservation and Utilization of Resources, held at Lake Success, New York, during August and September 1949.[20] The changes in the nature of the habitable world are plain enough, even in the formal language of the first paragraph of the resolution prepared by the Economic and Social Council in March 1947, which recognized "the importance of the world's natural resources, particularly due to the drain of the war on such resources, and their importance to the reconstruction of devastated areas, and recognizing further the need for continuous development and widespread application of the techniques of resource conservation and utilization."[21] It is not my intention to summarize the reports of this conference, as this is readily available elsewhere. More pertinent to our discussion is a review of some of the more general ideas expressed in that forum, some contradictory, some

---

20. *Proceedings of the United Nations Scientific Conference on the Conservation and Utilization of Resources, 17 August–6 September 1949, Lake Success, New York* (United Nations, Department of Economic Affairs, 1950) (hereafter cited as *UNSCCUR Proceedings*).

21. *UNSCCUR Proceedings*, Plenary meetings, vol. 1.

complementary, but all illustrating the divergences in emphasis that lack of real information accentuates.

First there was the thesis of Fairfield Osborn, author of *Our Plundered Planet*, that the astounding technological developments of the past century, together with an equally remarkable doubling of the world's population, had as one consequence the occupation of the habitable and cultivatable regions of the Earth, "leaving certain tropical regions and arctic regions as the last remaining frontiers."[22] Another consequence "has been that the drain upon the earth's resources has increased not upon a mathematical scale related to population growth, but upon a geometrical scale related to greater numbers of people demanding a greater variety of products from an infinitely more complex industrial system. This demand is still on the upward spiral, and will continue so as we strive for so-called higher standards of living for greater numbers of people." For the inorganic resources, the irreplaceable minerals, Osborn found in the technologist "our brightest, indeed our principal hope" in his ability to discover new resources or to devise substitutes. The situation regarding renewable resources, and the perennial productivity of the earth, is different. "One can even dare to make a further statement regarding the present situation, namely that if correct practices were now being generally applied to world forests, agricultural lands, and marine resources, coupled with adequate methods of distribution, the present haunting fears regarding adequacy of food supplies and other organic materials would no longer be justified." This is an interesting statement since it seems to imply that the problem is primarily one of conservation rather than a combination of conservation and population increase.

Mankind's record of husbanding the world's resources, Osborn argued, has been largely a failure; much of the damage has been of a permanent character, and these failures are being perpetuated in the present in most parts of the world. He argued that the failures of the past were not necessarily due to ignorance but were perpetrated "in the face of contemporary knowledge of their consequences." Moreover, he argued, "What we are seeking is the acceptance of a clear concept regarding man's relationship to his environment." This last sentence shows a wider gulf with the past than ordinarily would come from a general statement. If this problem were posed to the geographers, sociologists, and anthropologists of the latter part of the nineteenth century when so many of these earthshaking changes were progressing, the answer from 90 percent of them, I am confident, would have been that man is molded by his environment, that the physical environment, in conjunction

---

22. Fairfield Osborn, *Our Plundered Planet* (London: Faber and Faber, 1948). Subsequent quotations in the text are from this edition. [Along with William Vogt's *Road to Survival*, this book relaunched environmental thinking in the postwar era.—Ed.]

with some cultural factors, has determined human history and the rise and fall of civilizations. There is no doubt about it: the concept of the physical environment has undergone a radical change in emphasis, and I think it is to be explained as the change from static notions (such as mountains and rivers) to the idea of the environment as a complex interrelationship.

This was one point of view. Colin Clark, the Australian economist, presented another, with a widely different interpretation.[23] He took as his point of departure two quotations, one from William Vogt and the other from the *Economist*. Vogt had said that conservation was useless without a check on the world's population, and that fifty years hence most of the three billion people on Earth could be supported only at a coolie level. The *Economist* found that soil erosion "is no longer the menace it once was. . . . The application to the whole world of the farming standards that now prevail in the most efficient countries could probably produce a doubling of the food supply. . . . Before very long the normal state of affairs will return and the supply of the food in the world market will show a chronic tendency to outrun the effective demand for it." As Clark said, "it is not often that we have the spectacle of two authorities contradicting each other quite so categorically." Clark disagreed with both: while conceding Vogt's arguments for conservation, he denied that the world would never be able to support three billion or even more people. Vogt "has neglected or played down the possibilities of improvements in the technique of agriculture." The *Economist*, he said, "overrates the possibilities and neglects the time and effort necessary to bring them about." There is another factor that the *Economist* did not mention and that in the long run may be almost as important as improvement in agricultural technique, namely, the transfer of agricultural population by migration from overcrowded to fertile but underpopulated lands. But this will be an even slower and more difficult process.[24]

Clark's argument was that the problem is not one of shrinking land but of a declining rural labor force, and that the world's population can be supported at present and in the future through worldwide mobility of agricultural labor and a higher standard of living for this group. The steps in this argument are these:

1. After a painstaking analysis of world's population by countries, Clark concluded that the Earth's population is increasing at about 1 percent per year (that is, by about 20 million people).

---

23. Colin Clark (1905–1989) was a British (and Australian) statistician, economist, and public servant who was among the key early protagonists of the concept of GDP. At the time of the conference, Clark was director, Bureau of Industry, and under-secretary of state, Department of Labour and Industry, Brisbane, Queensland.—Ed.

24. Colin Clark, "World Resources and World Population," in *UNSCCUR Proceedings*.

2. The real quantity of farm products produced per man-hour of labor can, however, increase at the rate of 1½ percent a year. However, he added four important qualifications about this analysis:
   a. The demand for food per head is not constant but increases with increasing real income and standards of living.
   b. The 1½ percent per year rate of improvement would not hold if an increased agricultural population were densely crowded into a limited area.
   c. Farmers and farmworkers in the future can expect to work shorter hours and take longer holidays, in the manner of the urban population; thereby reducing production.
   d. The whole comparison rests on the assumption that the farm population remains constant, whereas in fact, throughout a large part of the world it is in rapid decline.

Clark ascribed this "flight from the land" to cultural and psychological factors, the breakdown of distinctions between rural and urban living, and increased social mobility. Technical improvements, he argued, will be insufficient to maintain the world's population if agricultural labor continues to decline. But even this decline can be offset—by a growth in rural populations of unexplored areas through geographic redistribution, and by a decline in the population of densely populated areas.

This, however, poses a big question. If there is this redistribution of the world's rural peoples, where will they go? The principal areas of unused cultivatable land in the world lie in Africa and South America, and the development of these would have to hold the key. For measuring the extent of farm products production for 1960 on a worldwide basis, Clark adopted as his areal unit "standard farm land," which he defined as land that is climatically fit for agriculture, without taking into account soil topography. Like Penck's estimate, it is based on a classification of climate, with, it must be confessed, no other justification than that climate alone is a good indicator of arable and cultivatable land. Clark's method of computation is quite simple: tropical land with regular (i.e., permanent, year-round) rainfall is given a value of 2 standard units, on the grounds that it can possibly produce two crops per year. Standard farmlands, with the value of 1, are in wet or humid areas, and in subtropical and temperate climates. A substantial area is counted as 5/6, 2/3, or ½ of a standard unit, based on the quantity and quality of rainfall. Semiarid or pasturelands are given a value of 1/100. Irrigated lands in hot climates are counted as 2 units.

It is obvious that by such methods, the amount of standard farmland will be very great indeed. Clark's figures add up to 61,863 square kilometers of standard farmland. While insufficient detail is given to see precisely the areal distribution, it is significant that out of 23,647 square kilometers in North and South America, 17,191 square kilometers are assigned to Latin America exclusive of Argentina; and that Africa, excluding Egypt and South Africa, is allotted almost 24 percent of the world's total standard farm land. These computations give, for the world, the equivalent of 15,286,347,300 acres of standard farm land. This does not, however, mean that this is the actual acreage of the world's farmland because of the various multipliers used, especially when the hot and rainy regions have all been multiplied by two.

After recovering from the astonishment that the farmlands of the world could be estimated on the basis of climate, even within such a minor error of 100 percent, it is apparent that Dr. Clark's hopes for a future 1½ percent annual growth in food against an annual population growth of 1 percent rests on three propositions: (1) a higher standard of living to maintain a high population in agriculture (a possibility); (2) a redistribution of the world's food producers either by migration or the growth of indigenous populations in the thinly settled areas capable of carrying on an efficient agriculture, both of which are very unlikely in the future; and (3) the intensive agricultural development of Africa and Latin America, a highly controversial subject. The lack of even a common meeting ground between Professor Clark's assumptions and students of these areas, such as Pierre Gourou, author of *Les pays tropicaux,* and Jean-Paul Harroy,[25] author of *Afrique, terre qui meurt,* is fully as striking as the juxtaposition of William Vogt and the London *Economist*. I cite Dr. Clark's analysis not in disparagement but to show again the exceedingly gross assumptions on which such estimates still rest. Methodologically, this is no advance over Ravenstein's estimate of fifty years ago. But more important still: when all the qualifications are considered, the range between the optimistic and the pessimistic outlooks seems quite narrow. In these matters it often seems that it is not the *content* but the *manner of writing* that establishes in the mind of the reader an optimistic or a pessimistic point of view.

In the discussion that followed, Dr. Clark admitted that he had made several assumptions, including attempting to use a climatic approach to make a crude ascertainment of the areas of land potentially available for agriculture; that he was not a statistician and not a soil scientist, and therefore could not engage in soil classification even if he had the data; that there was a very large margin of error—even a 100 percent error factor; and that in Africa and South America there are still vast

---

25. Pierre Gourou, *Les pays tropicaux: Principes d'une géographie humaine et économique* (Paris: PUF, 1947); Jean-Paul Harroy, *Afrique, terre qui meurt: La dégradation des sols africains sous l'influence de la colonisation* (Brussels: M. Hayez, 1949). [Glacken remarks on the side, "note the title again!"—Ed.]

areas of the unused land. However, his underlying message was that the future well-being of the world would depend on the solution of the problem of tropical soils. He said that the misuse of the tropical soils was a terrifying but definitely possible prospect, with the consequence that the world might become permanently hungered and impoverished.[26]

Many other topics surfaced during the conference. One of them concerned a disparity in the distribution of natural resources and access to modern technology. Speakers pointed out that owing to this uneven distribution of resources, the gap between the living standards of the more developed and the less developed areas in the world is widening. For example, it was pointed out that only one-fifth of the Earth's population lives in highly industrialized countries; that agriculture tends to be less highly developed in the so-called agricultural regions of the world; and that, within the relatively small areas of the world that are industrialized, great differences exist in the degree of development in industry and agriculture. Some argued that the solution to the problem was to be found in trade between industrialized and underdeveloped areas. Such solutions were based on the possibilities of achieving (1) a worldwide trading economy and (2) intentional agreement as to basic needs. The alternative, it would seem, is a continuance of the older trends.

Another issue concerned the relationship of the world's forests to civilization. The disappearance of forests has been associated with the development of civilization, as we have seen from the early statements of Hume and the later observations of Marsh and Ratzel. In our own times, several different relationships have received special emphasis: (1) a high use of forest products is associated with the most technologically advanced civilizations; (2) forest products can be used as effective substitutes for nonrenewable resources; (3) conservation and afforestation are needed not only to maintain the supply of forest products but to prevent erosion and floods; and (4) the distribution of the use of forest products is unequal. It has been said that one-third of the human race (in Europe, North America, the USSR, and Oceania) consumes 83 percent of the world's saw timber and 95 percent of the wood pulp. Worldwide afforestation on a gigantic scale has in fact been advanced as an indirect means of restoring the habitability of large areas of the Earth's surface. It has been estimated that there are three billion acres of forest now in use, with an additional five billion acres of virgin forest in Africa, South America, Alaska, and Siberia, and that one billion acres (of the four billion destroyed by man) might be replanted to forest, especially since much of this land

---

26. There is a fascinating exchange between Osborn and Clark in the aftermath of his presentation.—Ed.

exists where the population is most greatly in need of forest products. This gives a potential nine billion acres that, even if reduced to more conservative estimates, could provide enough wood fiber and its derivatives to help ameliorate world hunger and want. Here again, with the existence of scientific and technical skill, the possibility of hope is held out, but underlying this estimate is the assumption that the forests would become an international resource, used as renewable crops and not devastated.

Some conference participants regarded minerals and metals as more crucial than food, in the sense that the technological civilization required to produce a high standard of living is dependent on minerals and metals. Fear of the exhaustion of coal, petroleum, and the rarer elements, such as potassium and cobalt, it was pointed out, has persisted since the latter part of the nineteenth century. Indeed, the quantity of mineral products consumed between 1900 and 1949 far exceeded that of the whole preceding period of man's existence on Earth.[27] Some conference participants argued that the combination of increasing world population and the almost universal demand for a high standard of living (meaning the increased use of mineral products) would "place a strain on the metal resources which will almost certainly in the end be beyond the capacity of man and the nature is supply."[28] There was hope in the scientific discovery of new ores, in the improvement of techniques of extraction and processing, in conservation and the use of substitutes. Hope was also expressed that plastics and other synthetic products "will eventually develop into a final substitute."[29] The danger, however, is that these ideas are purely speculative, hopes, not promises.

A great deal of the subsequent discussion at the conference concerned the need for the deployment of scientific and technological techniques to map the distribution of existing resources and conditions, engage in conservation efforts, and explore technological substitutions. However, some broader, more philosophical issues also emerged. One of them concerned the tendency, in the modern literature on world population and world resources, to generalize on a global basis. The commonest form of this approach, discussed earlier, entails computing the world's population and dividing it into the estimated number of acres or hectares of the world's land to get a quotient of so many units of land per inhabitant of the globe. This approach was, however, challenged forcibly by Professor John D. Black, who

---

27. Hugh Keenleyside, "Critical Mineral Shortages," in *UNSCCUR Proceedings*, JI: 37–45. [Keenleyside was deputy minister of the Department of Mines and Resources in Canada, and also founder and member of a number of organizations for the study of Arctic and Canadian geographic conditions.—Ed.]

28. Ibid.

29. Ibid.

argued that it assumes that "the population of the earth were one vast drove of hogs feeding out of a common trough."[30]

In my own view, since the end of World War II the seeming contradiction between world total resources and world population and total resources and the population of each area or nation is not so clear-cut as might be supposed. Certainly the total world population and the estimated total resources in food or minerals are significant figures. Statistics of this type will always be compiled and will always be used. They are bound to be of great significance in the hands of those who attempt to discover major trends in the world, or who try to find a meaning in history. A worldview of one sort or another is at the basis of most optimistic or pessimistic appraisals. There is, as Clark pointed out, no justification for using these as a working figure expressing the relation of all mankind to the resources of the environment. Any such approach assumes fluidity in human affairs, an international or a worldwide economic system, and negotiations in a peaceful atmosphere.

World totals are important in that they emphasize relationships that exist, despite the walling off of large areas of the world from one another. They emphasize it in this way: the self-containment of the standard of living of industrialized nations, especially since World War II, means the responsibility of each nation to take care of its own population has broken down in the face of world conditions. These conditions have the effect of leveling off the standard of living of the better-fed nations. The experience of the United States is an example. Its wealth was spent not only in the prosecution of the war but also in relief measures, in conjunction with other nations, after the war. Following the realignment of the free world versus the communist world, both military and relief measures have been necessary. It is inconceivable that in the further conduct of this conflict, to offset communist propaganda, immense sums will be spent for nonmilitary purposes, to achieve or guarantee success. Our own era is an illustration of a declining standard of living, a redistribution of resources not so much in response to world hunger directly but to political ideologies, of which large-scale hunger is a part, but only a part.

Probably from no other single source could one find the ideas that in our time are shaping our concept of the nature of the habitable world. Perhaps the foremost idea is that it is possible, with important qualifications, such as the maintenance of peace and international cooperation, for science and technology to provide both for the means of repairing the worldwide devastation—an expression also of primitive and civilized societies in which science has equally contributed—and at the same time to deal affirmatively with the expected increases in the world popu-

---

30. John D. Black, "Economic Considerations in Conservation and Development," in *UNSCCUR Proceedings*. [Black was a professor at Harvard University.—Ed.]

lation. The world of the twentieth century appears to be a little explored one. The areas of hope lie in technology and science; the areas of urgency lie in the lack of technicians for the soils and for resource conservation throughout the world. International barriers of all sorts, differing from country to country, impede the technological possibilities. To me, we have now reached a conception of interdependence of such delicacy, even fragility, that its achievement would demand the most favorable historical circumstances. So much depends on forest conservation, if we may paraphrase again; so much depends on the proper care of tropical soils, so much depends on mineral and metal conservation, on intelligent resource appraisal, on reconciling human institutions with the stability of the environment; so much depends on a world labor supply, on an adequate system of land classification, that we may fairly ask whether the solution of these problems is of the same order of difficulty as the abolition of war.

This series of discussions based on various points of view from the early seventeenth to the middle of the twentieth centuries is open to the grave criticism that they may be haphazardly or arbitrarily chosen, and that they too represent a high degree of selectivity in choice. But I believe that an exhaustive sourcebook of such materials would show little more. In essence, there has been a great transformation in men's conceptions of their relation to their physical environment. This has been complicated by the immense social questions that have followed the great increases in world population since 1750. And the critical abandonment of the idea of inevitable progress has removed the hope of automatic corrections with time.

A Selected Bibliography of Clarence Glacken's Works

Books

The Great Loochoo: A Study of Okinawan Village Life. Berkeley: University of California Press, 1955.

Traces on the Rhodian Shore: Nature and Culture in Western Thought from Ancient Times to the End of the Eighteenth Century. Berkeley: University of California Press, 1967.

Journal Articles and Book Chapters

Control of Climate in the Home. *Geographical Review* 41 (1951): 164–65.

Changing Ideas of the Habitable World. In *Man's Role in Changing the Face of the Earth*, ed. W. L. Thomas, 1236. Chicago: University of Chicago Press, 1956.

The Origins of Conservation Philosophy. *Journal of Soil and Water Conservation*, 1956.

Man and Earth. *Landscape: Magazine of Human Geography*, 1956, 63–66.

Culture and the Idea of Nature. *Yearbook of the Association of Pacific Coast Geographers* 18 (1956): 23–27.

Count Buffon on Cultural Changes of the Physical Environment. *Annals of the Association of American Geographers* 50 (1960): 1–22.

This Growing Second World within the World of Nature. In *Man's Place in the Island Ecosystem*, ed. F. R. Fosberg. Honolulu: Bishop Museum Press, 1963.

Introduction. In *The Geographical Lore of the Time of the Crusade*, ed. K. W. Wright. New York: Dover Publications, 1965.

Reflections on the Man-Nature Theme as a Subject for Study. In *Future Environment of North America: Transformation of a Continent,* ed. F. F. Darling and J. P. Milton, 767. New York: Natural History Press, 1966.

On Chateaubriand's Journey from Paris to Jerusalem, 1806–1807. In *The Terraqueous Globe.* Los Angeles: A. Clark Memorial Library, University of California.

Man against Nature: An Outmoded Concept. In *The Environmental Crisis,* ed. H. W. Heinrich, 127–42. New Haven, CT: Yale University Press, 1970.

Man and Nature in Recent Western Thought. In *This Little Planet,* ed. M. Hamilton, 163–201. New York: Scribner's, 1970.

Environment and Culture. In *Dictionary of the History of Ideas: Studies of Selected Pivotal Ideas,* ed. Philip P. Wiener, 4 vols., 2:127–34. New York: Charles Scribner's Sons, 1973.

Culture and Environment in Western Civilization during the Nineteenth Century. In *Envi-*

ronmental History: Critical Issues in Comparative Perspective, ed. Kendall E. Bailes. Lanham, MD: University Press of America for Adirondack Museum, 1985.

Reflections on the History of Western Attitudes. *GeoJournal* 26 (1992): 103–12.

Book Reviews

Review: A Population History of the World. *Journal of Economic History* 11 (1948): 74–75.

Review: Heinrich Schmitthenner, Studien über Carl Ritter. *Geographical Review* 43 (1953): 288–89.

Review: Splendour of Earth: An Anthology of Travel. *Geographical Review,* 46 (1956): 135–36.

Review: Jeannette Mirsky, Elisha Kent Kane and the Seafaring Frontier. *Geographical Review* 46 (1956): 137–38.

Review: The Reason of State and The Greatness of Cities. *Landscape* 6 (1957): 36–37.

Review: A Volume of Travel in Space and Time which have Delighted, Intrigued, and Intimidated Man, Geoffrey Gringson and Charles H. Gibbs-Smith. *Geographical Review* 47 (1957): 285–86.

Review: George Perkins March, Versatile Vermonter, David Lowenthal. *Geographical Review* 49 (1959): 437–48.

Review Essay: The City in History, Lewis Mumford. *Landscape* 11 (1961): 33–35.

Review: Friedrich Ratzel, a Biographical Memoir and Bibliography, Harriet Wanklyn. *Geographical Review* 52 (1962): 467–68.

Review: The Domesday Geography of Northern England. *Geographical Review* 54 (1964): 285–87.

Other Publications

Obituary: William Vogt (1902–1968). *Geographical Review* 59 (1969): 294–295.

A Late Arrival in Academia. In *The Practice of Geography,* ed. A. Buttimer, 20–34. London: Longmans, 1983.

Index

Acosta, José de, 29
adaptation, concept of, 51, 55, 59, 66, 140, 147, 151, 168, 174
afforestation, 193. *See also* deforestation
Agassiz, Louis, 54
age of discovery, 37, 46, 92, 103, 110, 156. *See also* Humboldt, Alexander von
Age of Reason, 1, 3, 12
agriculture: agricultural peoples, 10; agricultural revolution, 186; lands, 179, 189; shifting cultivation in, 29, 178; study of, 133, 135
Albert the Great, 68
Alembert, Jean le Rond d', 4
Alsberg, Carl S., 170–71
animal kingdom, 64, 73, 81–82; spiritual, 62
animals, 8–12, 16–17, 37–38, 48–49, 52, 55–57, 63–64, 66–70, 73, 85–86, 150 (*see also under individual names*); geographic distribution of, 56
Ansichten der Natur (Humboldt), 47–48
anthropogenic environmental change, 139–41
Anthropogeographie (Ratzel), 103, 106
anthropopsychism, as "man-Soulism," 60
apes, 48, 75–76, 78–79, 85; Anthropoid, 86 (*see also* chimpanzees); humans and, 52
Argyll, Duke of (George John Douglas Campbell), 54, 58–62
Aristotle, 35, 73, 180, 183
artificial: fertilizers, 109, 115, 124; selection, 57, 63
Aspects of Nature (Humboldt), 26

Bacon, Francis, 53–54
Ballod, Carl (Kārlis Balodis), 114–18, 122, 126; on population growth, 116
Banks, Joseph, 37, 62
barbarism, 88, 94, 100
Barrow, John, 15
beauty, concept of, 73
Bell, C., 77
Bellod, Carl, 95
Bennett, H. H., 172
Bibb, Cyril, 80
Biese, Alfred, 32
birth control, 96, 125–26, 185
Blache, Paul Vidal de la, 152
Bodin, Jean, 68, 140, 183
Bonpland, Aimé, 22, 39, 48
Boussingault, Jean-Baptiste, 102
Bowman, Isaiah, 166, 168
"Bridgewater Treatises," 61
Bruhnes, Jean, 152
Bryce, James, 140
Buckle, Henry, xxiv, 28, 140, 161–62
Buffon, comte de (Georges-Louis Leclerc), xxii–xxiii, 21, 24–26, 28, 30, 37–39, 41, 52, 62–63, 73, 91
"Burial of Olympia, The" (Huntington), 149
Butler, Joseph, 53–54

Campbell, George John Douglas (Duke of Argyll), 54, 58–62
Candolle, Alphonse de, 74, 150
carnivores, 8–9, 63, 81. *See also* animal kingdom

carrying capacity, concept of, 95
"cats-to-clover chain," 70
Chateaubriand, François René de, 21, 26, 37–39, 41
chimpanzees, 79
Cicero, 36
civilization, 3, 11, 25, 36, 78, 144, 146, 157–59, 162–63, 174, 177–79; agriculture and, 186; Buckle and, 28; in China, 158; conservation and, 176; Darwin's analysis of, 78; environmental effect on, 108; fertile plains and, 9; forest and, 193; forests to, 193; genesis of, 162–63; high, 88, 118, 144; Humboldt on, 45, 128; Huxley and, 83; Jacks and Whyte on, 173; man and, 10, 78, 82; Mayan, 163; migrant of 1937 and, 168; nature and, 41; population and, 124; products of, 93; soil erosion and, 173–74; Sumeric and Minoan, 163; technological civilization, 194; transformation and, 100; Wallace and, 85, 87, 93–94; Western, 40–41, 62, 68, 75, 90
Civil War, U.S., 102. See also wars and warfare
Clark, Colin, 167, 190–92, 195
climate, 14, 28–29, 68–70, 116–17, 123–24, 127–29, 141–42, 144–47, 149–51, 160–62, 191–92; classification of, xxv, 126–27, 150–51, 191; Darwin and, 56, 68–69; Humboldt and, 28; species and, 69; Virey on, 10–11
climatic change, 69, 99, 141–42, 144–47, 153, 161–62; and population history, 161
climatic classification, 127, 151
climatology, 107, 128, 135; Ballod on, 116
colonization, 46, 123, 137, 146, 155, 166, 168, 170
Condorcet, marquis de, 1, 6
conservation, xxiii–xxiv, 6, 12, 101, 104, 141, 172–73, 175–76, 178–79, 188–90, 193–96; movement, 178; resource, 188, 196; Theodore Roosevelt's committee on, 172; Vogt on, 190
Cook, Captain, 29, 37
cosmopolitanism, 21, 47
creation of God, 79; nature as, 55
crime, 94, 100
Crookes, William, 109
cultivation, 15, 18, 21, 33, 43, 104, 110–11, 116, 124, 127, 174–76
Curtius, Ernst Robert, 139

customs, 11, 34, 157, 180
Cuvier, Georges, 54

Darwin, Charles, xxiii, xxv–xxvi, 22, 26, 28, 32, 49–59, 61–78, 80, 89–91, 107–8; ecological implications and theory of, 65–75; *Philosophy of Nature*, 52–65; and systems of classification, 77; theories, 55–57, 62, 79, 83, 91, 108, 135, 138, 183; theory of evolution of, 60, 62
Darwin, Francis, 61
Davis, Jefferson, 98
Day of Judgment, 187
deforestation, 95–96, 101–2, 104, 139, 142, 145, 163, 174. See also afforestation; forests
democracy, 12
Descent of Man, The (Darwin), 52, 65
designed earth, 137–39
Desmolins, 162
despotism, 1, 4, 6, 11–12
development, theory of, 61
Duerst, Johann Ulrich, 144
Dust Bowl, the, xxi, 166, 172, 178–79
dust storms (1934), 173–74. See also Dust Bowl

"Earth Answers Back, The" (Vogt), 188
Earth, capacity of, 114; and Penck, 132
East, Edward M., 93, 95, 97, 118–26, 174
East, land estimation of, 121
ecological relationships, 56, 65, 72, 187
ecology, xxv–xxvi, 47, 50–52, 55, 71, 102, 135–36, 166, 181, 183; and evolution, 71
economic nationalism, 173, 175, 178
economic system, 195
Ehrlich, Paul, 2
Elements of Geology (Lyell), 84
Engels, Friedrich, 83
Ense, Varnhagen von, 47
environmental theories, 56, 96, 139–65, 188; trends in 133–41
environmental: determinism, xxii, 133, 146–47, 164, 183, 188; influences, xxii, 79, 145, 147, 149, 154–55, 159–61, 188
epidemics, 11, 68–69, 123
equilibrium, 8, 10–12, 62, 119, 125, 175, 182–83
erosion, 12, 101, 126, 174–77, 179, 193
Essay on the Principle of Population, An (Malthus), 12–13, 16, 138

European: civilization, 43, 92, 164 (*see also* civilization); culture, 43, 95, 108, 164
Europeanization, 104, 106
Evidence as to Man's Place in Nature (Huxley), 78
evolution, xxvii, 32, 49, 52, 56–57, 62–63, 65–66, 69–71, 79, 83–84, 86 (*see also under* Darwin); Huxley on, 80; by means of natural selection, 57; theory of, 22, 49, 51–52, 60, 62, 65, 71, 73, 79, 83–84, 86–87
existence of God, 90
exploitation, 11, 98, 169–70, 178, 182–83, 187
explorations, 21, 44, 110, 116, 137, 146, 153–54
extinction, 67, 76, 100; of undeveloped populations, 92

famine, 10–11. *See also* poverty
farm land, 191–92. *See also under* agricultural; soil
Febvre, Lucien, 151–52
fertility, 9, 29, 116, 118, 124–25, 132, 178
"Fertilization of Orchids" (Darwin), 59. *See also* Darwin, Charles
feudalism, 99, 177–78
Fircks, Arthur Freiherr von, 114–15, 126
fire, discovery of, 76
Fischer, Alois, 96, 129–31
Fiske, James, 139
floods, 104, 163, 172, 193
Fontenelle, Bernard Le Bovier de, 4
food, 7–8, 67–69, 113–14, 117–18, 122–24, 127–29, 179, 190–92; for animals, 124; chain, 73; production, 116, 120–21, 124, 171
forests, 9, 30, 35, 48–49, 100–101, 104, 111–12, 141, 154, 186–87, 193–94; conservation of, 104, 196; destruction of, 104; disappearance of, 193
Forster, Georg, 29, 37, 38, 39
Forster, Johann, 29, 37
Franklin, Benjamin, 16, 182
French Revolution, xxv, 7, 155

Geoffroy Saint-Hilaire, Étienne, 62
geographic distribution, concept of, 24, 51, 73, 75, 151–52
Geographical Distribution of Animal, The (Wallace), 74
Géographie botanique raisonée (Candolle), 74

Ghiselin, Michael T., 51, 53, 61
Glacken, Clarence, xix–xxvi, 21; and Darwin's idea of nature, 51; Hoosen on, 21
Glacken, Mildred, xxiii
Glacken Archives, xxv
global overpopulation, 115
glory-of-God theories, 56
Godwin, William, xxvi, 2–3, 5–6, 12–20, 108, 114, 138, 167
Goethe, Johann Wolfgang von, 26, 37, 47
Golen, 183
Goodrich, Carter, 172
gorillas, 79
Graham, E. H., 183
Gray, Asa, 53, 62
Great Depression, the, xxi, 171, 179
Guyot, Jules, 158

habitable earth, x, 11, 20, 101, 103, 129, 132, 169
habitable world, x, xxi–xxii, xxv–xxvii, 17–18, 102–3, 106–8, 123–26, 128–29, 180, 187–88; Alsberg's appraisal of, 171; descriptions of, 107
Haeckel, Ernst, 26, 47, 71
Halloy, Jean Baptiste d', 158
harmony, 90, 99, 135, 188; Humboldt on unity and, 23, 25–26; Ratzel on, 105; Virey on, 1, 8–10, 12; Vogt on, 182
Harper, Floyd A., 180
Harroy, Jean-Paul, 192
"*Hauptmomente*" (Humboldt), 45
Hedin, Sven Anders, 145
Hegel, G. W. F., 62, 155–58
Hehn, Victor, 150
Helmholt, H. F., 140
Herder, Johann Gottfried von, 49, 103, 108, 160–61, 183
Herodotus, 98, 163, 183
Hilgard, Eugene W., 135
Hippocrates, 68, 98, 140, 162, 183
Hodson, Margaret, xx, 62
Hogarth, David George, 145
Holde, Dr., 15
Holstein, Penck Fischer, 129–31
Holstein, W., 129
Hooker, J. D., 53, 63
Hoosen, David, xxv, 21
Hudson, W. H., 36, 71

INDEX 201

human: agency, xxii–xxiii, 39, 52, 76, 81, 95, 108, 139, 141–42, 145–46, 150; cultures, xxi, xxv, 91, 133, 136, 151, 165, 172, 178, 188; generations, 10; geography, 34, 42, 103, 148, 159
human race, 79, 82, 90–91, 124; and creation of God, 55, 75, 91; future of, 132; Humboldt on, 24; Huxley and, 80; Judeo-Christian belief on, 55, 75, 91; primitives and, 87; Virey on, 11
humanism, 21; and Humboldt, 47; Pierre and, 26
Humboldt, Alexander von, xxv, xxvii, 9, 21–50, 52–53, 55, 62, 73, 91, 118, 136; and landscape, 27; and nature, 29; in plant geography, 136; on role of Buddhism, 42
Hume, David, xxii, 161, 193
Huntington, Ellsworth, 118, 122, 125–26, 128, 143–49, 152–54, 161–63
Huxley, Thomas Henry, xxv, xxvii, 26, 51, 53, 63, 75, 77–83, 85, 90–91, 100; and concept of man, 81; and Darwin's theory, 79; on evolution, 80

imperialism, 66, 96
industrial: civilization, 92 (*see also* civilization); revolution, 94, 169, 186
industrialization, 102, 186
interspecies competition, 68
isolation, 74–75, 104

Jacks, G. V., 166, 173–78
James, William, 139
Jevons, William Stanley, 109, 113

Kant, Immanuel, 77
Keller, Albert Galloway, 157
Keynes, John Maynard, 82, 119
Kirchhoff, Alfred, 152
Klaproth, Julius von, 155
Köppen, Wladimir, 127–29, 150
Korzybski, Alfred Habdank Skarbek, 184–85
Kosmos (Humboldt), 21, 23–24, 31–32, 34, 36, 40, 47, 49
Kroeber, Alfred, 149
Kropotkin, Prince Peter, 136, 142–44, 153–54

laissez-faire capitalism, 78
Lamarck, Jean-Baptiste, xxiii, 22, 24–25, 30–31, 49, 52, 55, 57, 62, 67, 71

land: distribution, 117; per inhabitant, 194; settlement, 167–72; suitable for agriculture, 170–71
laws of matter, 47
Lebensraum, concept of, 95
Leclerc, Georges-Louis. *See* Buffon, Comte de
Lefebvre, Henri, 161
Lewis, Meriwether, xxii; and Clark Expedition, 17
Liebig, Justus von, xxvi, 19, 102, 109, 124, 126, 134, 181
Limits of Land Settlement (Bowman, ed.), 166–68
Lincoln, Abraham, 97
Linnaeus, Carl, 7, 22, 62, 75–76
List, Georg Friedrich, 128
Lovejoy, Arthur, xxvii, 24, 32
Lowdermilk, Walter, 172
Lowie, Robert, 149
Lucretius, 36, 63
Lyde, Lionel, 141
Lyell, Charles, xxvi, 53, 63, 76, 84, 139

Mackinder, Halford John, 152–55, 157
Malay Archipelago, The (Wallace), 84, 92–93
Malthus, Thomas: Darwin's acceptance of, 74; doctrine of, 83, 108, 125, 167, 183, 185, 187; doctrine of population, 14, 63, 76, 82; *Essay on Population*, xxii, 1–3, 9, 12–16, 18–19, 49, 63–64, 67–68, 82–83, 108, 185; theory of, xxvi, 15, 18, 50, 57, 82–83, 108, 119–20, 124, 139, 156–57
Malthusianism, xxvi, 5, 51, 66, 115, 117, 126, 157, 167, 185, 188
man: as *contemplator mundi*, 54; in nature, 11, 75, 79, 84, 90–91; origin of, 86; as peacemaker, 9; as social animal, 35; as supreme moderator, 10
Man and Nature (Glacken), xxv, 1, 52, 81, 91, 95–97, 152, 166, 185
"man's place in nature," 75–94
Marbut, C. F., 130
Marsh, George Perkins, xxiv, xxvii, 52, 95–102; on Liebig, 102; on nature, 100
Marshall, Alfred, xxiii–xxiv, xxvii, 52, 91, 95–102, 104, 108, 113–14, 126, 141, 172
Marx, Karl, 83, 164
McNeil, J. R., xix
mechanical invention, 120–21, 125

mechanization of industry, East on, 120. *See also* industrialization; industrial revolution
Merz, John T., 32
meteorology, 24, 99, 107
Michelet, Jules, 139
migrations, 1, 10–11, 76, 144, 148, 154, 168, 192; of agricultural population, 190; barbarian, 142; to California, xxi; and civilization, 146; and Europe, 142, 166; and food production, 137; Carter Goodrich and, 172; international, 121, 123, 137, 140; nomadic, 142, 163; and population growth, 17; Sauer on, 170; of social plants, 23
Mill, John Stuart, 94
Montesquieu, xxii, 68–69, 140, 160–61, 183
multiplication, law of, 87
Murray, James Augustus Henry, 41, 61, 97
Murray, John, 41, 61
Myres, J. L., 145

Napoleon, 1, 3, 12; wars of, 7. *See also* wars and warfare
nationalism, 172, 175, 178, 183
natural history, xxii, xxvi, 1, 7, 9, 12, 26–27, 55, 61–64, 66, 73; golden age of, 37
naturalists, 60, 71
"natural religion," 47
natural resources, 102, 171, 175, 188, 193
natural selection, 55–58, 62–63, 65–66, 69, 71–74, 76–78, 83–89, 91, 93, 139
natural theologians, xxiii, 53, 55–57, 63, 77, 91
nature: concept of, 22, 25–26, 29, 31, 35, 41, 63; constancy of, 53–54; economy of, 51–52, 55–56, 66–67, 71; and Humboldt, 30; of man, 84, 87 (*see also under* man); teleology in, 62
Naturphilosophie, 27, 30–31
Newman, H. 70n44

Oligochaeta, 102
organic nature: destruction of, 108; Virey on, 7
organisms, 55–57, 65, 107, 136, 187
Origin of Species, The (Darwin), 32, 49, 51–53, 55, 57, 60, 62–65, 73–75, 79–80, 100, 138
Osborn, Henry Fairfield, 180, 189
overpopulation, 11, 123, 142, 167, 179
"overruling intelligence," 87
Ovid, 36

Pallas, Peter Simon, 37
Partsch, Joseph, 103
Pearl, Raymond, 125
Pearson, Frank A., 180
Penck, Albrecht, 126–32, 191
Peschell, Oscar, 102
Petermann, August, 110; *Mitteilungen*, 137
plants, 10, 12, 16, 30, 37, 57, 63, 67, 69–70, 73, 136, 150, 152; cultivation of, 43; and flowers, 8; geographic distribution of, 56; morphology, 119
Playfair, John, 38, 63
Pliny, 141, 183
Pluche, Noël-Antoine, 37
Polybius, 98
population 5, 95, 109–10, 117, 122, 167, 179–80, 185, 189–91: balance of, 108; concentration of, 105; densities, 3, 105–6, 131, 168–70; Europe and problem of, 175; expansion of, 123, 169; pressure, 17, 82, 124, 170, 175–76; problems, 115, 169; redistribution of, 166, 171
population growth, xxi, 5, 7, 14, 17–18, 105, 113–14, 119–20, 129, 169, 181–82; epidemics and parasitism on, 68; rebellion against, 17
poverty, 2, 78, 93–94, 107, 155
predation, 63, 68, 73
Prentice, Ezra Parmalee, 180
"primitive" peoples, 45, 52, 76–77, 84–85, 87, 100, 140; extermination of, 92; oppression of, 66
"principle of population," 13–14, 16, 63–64, 76, 82–83
Principles of Geology (Lyell), 66, 84
Pumpelly, Raphael, 142–44; Turkestan expedition of, 142, 144

race, 11, 14, 64, 66, 75, 79–80, 82, 87–88, 90–91, 119–20, 123–24, 132; intermingling of, 17; of white, 120, 123–24, 130
Rape of the Earth, The (Jacks and Whyte), 166, 173
Ratzel, Friedrich, xxiv, xxvii, 95, 102–8, 110, 118, 126, 129, 140, 147–48, 162; Glacken on, 95; and concept of man and earth, 104
Ravenstein, E. G., 95, 108–15, 118–19, 126, 192
Ray, John, xxiii, 54, 62, 125
Reign of Law, The (Argyll), 58–59

INDEX   203

Richthofen, Ferdinand von, 102, 142
Ritter, Carl, 98, 103, 108, 139–40, 148, 152, 157–58, 183
Road to Survival (Vogt), 166, 179–80, 186
Rousseau, Jean-Jacques, 1, 3, 21, 26, 31, 38–39, 41
Ruskin, John, 60

Saint-Hilaire, Geoffroy, 62
Saint-Pierre, abbé de (Charles-Irénée Castel), xxvi, 1–6, 188
Saint-Pierre, Bernardin de, 22–23, 28, 31, 37, 41, 52, 63; *Études de la nature*, 27
sanitation, systems of, 187
Santa Barbara de Arichuna, mission of, 48
Sauer, Carl O., xx–xxiii, 166, 169–70, 172–73
"savages," 9, 17, 77, 81, 84–88, 93
Schiller, Friedrich, 34
Schouw, Joakim Frederik, 73
science: history of, 31, 44, 49; scientific ecology, 135–36; scientific exploration, 136–37
Sears, Paul, 172
Semple, Ellen Churchill, 147–48
Shaler, Nathaniel Southgate, xxiv, 182, 188
Shantz, H. L. R., 130
Simon, Julian, 2
Smyth, Albert Henry, 16
soil and land, xxv–xxvi, 8–9, 82, 99, 105–6, 114–16, 118, 124–28, 132–35, 172–78, 186–88; and civilizations, 172–78; conservation, 101, 173, 175, 178; distribution of, 135; East's concept of, 124; erosion, xxi, 101–2, 132, 145, 172–77, 179–81, 190; exhaustion/deterioration, 17, 104, 124, 126, 174; Jacks and Whyte on movement on, 178; productivity of, 176; study of, xxv, 133–35; tropical, 29, 118, 132, 135, 187, 193, 196
species, xxi, 23, 35, 44, 51–53, 62–65, 67–69, 72–75, 79–80, 100; reproduction and diseases, 10
Spengler, Oswald, 161–62
Stein, Marc Aurel, 145
Stoddart, David, xxv
Strabo, 98, 161
"Struggle for Existence in Human Society, The" (Huxley), 80
struggle for existence, notion of, 62, 65, 70, 80, 93
Styles, Dr., 16

Sumner, William Graham, 157
survival, laws of, 87
"survival of the fittest," 93

Taylor, Griffith, 97, 101, 162, 183
Teggart, Frederick, xx, 32, 92–93
temperate zones, 113, 115–16, 118, 128, 170
temperature, 24
theologians, natural. *See* natural theologians
Thirlmere, Rowland, 144
Thomson, Sir John Arthur, 71, 91
Thoreau, Henry David, 97
Thucydides, 183
Tibullus, 36
Tournefort, Joseph Pitton de, 23
Toynbee, Arnold Joseph, 144, 147, 161–63
Traces on the Rhodian Shore (Glacken), xix, xxi, xxiii, 23, 37, 51
trade, 46, 102, 128, 132, 137, 161, 185, 193
transformations, 39, 81, 83, 100
Treaty of Utrecht, 4
tropics: cultivation in the, 178; soils of the, 29, 118, 132, 135, 187, 193, 196. *See also* soil and land
Tull, Jethro, 134
Turner, Frederick Jackson, 153, 155–57
Tyndall, John, 61

UN Conference on Conservation and Utilization of Resources, 188–96
uniformitarianism, 63
Unity of Nature, The (Argyll), 58–59
universe, the, 7–8, 12, 25–26, 45, 47, 79–80, 88, 90; Humboldt and, 45
Utterström, Gustaf, 161

variation, laws of, 86
Variation of Animals and Plants under Domestication, The (Darwin), 52, 74
vegetable kingdoms, 64, 92. *See also* plants
Virchow, Rudolph Carl, 158
Virey, Julien-Joseph, xxvi, 1, 6–12, 188
Virgil, 36
Vogt, William, 166, 173, 180–88, 190, 192
Volney, C. F., 143
Voltaire, 1, 3, 161

Wagner, Hermann, 74–75, 127
Wagner, Moritz, 74, 102

Wallace, Alfred Russell, xxii–xxiii, xxv, xxvii, 26, 28, 50–53, 59, 74–75, 77–78, 80, 83–94
Wallace, Robert, 157
wars and warfare, 7, 63, 66, 119; abolition of, 6, 196; Franco-Prussian War, 102; Mexican War, 97. *See also* World War I; World War II
Warming, Eugene, 136
wealth accumulation, 94
Webb, Walter Prescott, 153, 156
Webster, Noah, 141
Whewell, William, 53–54
White, Gilbert, 27, 37, 71; *The Natural History of Selborne*, 65

Whyte, R. O., 166, 173–78
Woeikof, Alexander Ivanovich, 177, 182, 188
world history, xx, 140, 146, 152, 155, 173–74, 186
world trade, 113, 127, 160, 175
World War I, xxiv, 102, 108, 119, 123, 155–56, 171, 174, 179, 186
World War II, 123, 155, 172, 175, 178–80, 195
World's Hunger, The (Pearson and Harper), 180

"yellow peril," 119
yields, observations of, 134

Zittel, Karl Alfred von, 102

www.ingramcontent.com/pod-product-compliance
Lightning Source LLC
Chambersburg PA
CBHW061442300426
44114CB00014B/1796